AF541587

Psychology and Human Behaviour

PSYCHOLOGY AND HUMAN BEHAVIOUR

Edited by

K.C. Dubey

OMEGA PUBLICATIONS

NEW DELHI-110 002 (INDIA)

OMEGA PUBLICATIONS
4378/4-B JMD House, Ansari Road,
Daryaganj, New Delhi – 110 002
Phone: 65901906, 23278062 (Off.)
E-mail : omega_publications@yahoo.com

Head Office:
79/23, Laxmi Garden,
Near Satya Jyoti School,
Gurgaon (Haryana)

Psychology and Human Behaviour

New Edition 2017
ISBN 978-81-8455-132-7

PRINTED IN INDIA

Published by Mahendra Garg for Omega Publications, New Delhi -110 002 and Printed at Tarun Offset Press, New Delhi - 110 053

Preface

Many studies of adolescence, psychological, sociological, and anthropological, have been made through the years. Adolescence has been studied as a period of development, as a crisis period in the child's life and in his interpersonal and social relations, and also as a social problem confronted by each succeeding adult generation. Consequently, an overall "traditional" conception of the nature of adolescence has become quite thoroughly established. The following statement is an attempt to characterize that traditional conception.

As a result of physical growth changes, and changes in physiological functioning the child experiences a complex of psychological accompaniments—feelings, emotional stirrings, motivating "drives"—that are completely new and strange to him. He even feels strange to himself as a physical being. He is disturbed by the new quality of his emotions. He experiences an upsurge of erotic desires along with the resurgence of the old so-called Oedipus complex-his earlier erotic attachment to his mother and rivalry feelings toward his father. These feelings he finds inconsistent with the ideals and standards of right and wrong, which he had previously internalized, and so there is conflict. Anxieties, feelings of guilt, apprehension, resentment arise. He finds himself in a state of bewilderment, a state of "storm and stress."

The major topics dealt in this book are : *Growth of Human Changes; Psychology of Human Mind; Human Behaviour and Emotions; Role of Mental Health; Group Behaviour and Human*

Experimentation; Functional Development; Development of Maturity and Impacts; Behaviour and Social Interaction; Development of Mental Process; Process of Individuality; etc.

No doubt, these will serve the purpose of trainees and trainers, professionals and policy planners in the field. Since the sources of information are all secondary, we express our gratitude to the scholars whose works are cited or substantially made use of. We are thankful to all those who rendered ready help and cooperation while working on this project.

We express our gratitude to various scholars, teachers and friends for their assistance and guidance. Finally, we thank our publishers for bringing out this book in very limited time.

—Editor

Contents

1

Growth of Human Changes

As the young person approaches adulthood certain factors in his experience conspire to make him more consciously concerned about himself as a person. He becomes self-conscious in a way that he never was before. The glandular changes described above initiate feelings and emotional qualities that are new and strange to him. The physical changes that are rapidly taking place in body dimensions, in physiognomy, in genital organs, the beginning of the boy's beard growth and voice change and the swelling of the girl's breasts, all contribute to feelings of strangeness. And when the child almost suddenly, as it seems to him, finds that his reach is longer, his feet bump into things like they never did before, that things about him more than ever before get in his way and get knocked over, he begins to feel as if he is dealing with a different person than his former self which he was able to manage so expertly.

The early part of the adolescent period has sometimes been called the "awkward age." Many adolescents are self-conscious concerning their posture and gait. Some show a certain inhibition or hesitancy of movement as though they were holding their muscles in check and did not feel free to let themselves go. Self-consciousness with respect to height may express itself by sloping shoulders, and the short adolescent boy may try to compensate for his shortness by an upward tilt of his head, squaring of his shoulders, and a tendency to let his weight rest more on his toes than on his heels. Sometimes, in his bearing and by the character of his physical movement, the adolescent calls attention to physical features which he would actually like to conceal.

If there is such a thing as an awkward age, it is during this relatively brief period of puberal change. All of this strangeness, confusion, and lack of self-management contributes greatly to what is called the identity crisis. It is .he problem of reconciling present experiences, perplexities, and embarrassments with ones past and of orienting oneself toward a meaningful future.

There are, of course, wide individual differences in the seriousness and the disturbing aspects of this identity problem. Many young people appear not to be bothered in the least. They continue to be their outgoing selves, maintaining their usual easy relations with their peers, older persons, and with parents and family. Others, however, tend to "retreat into themselves," becoming affected or difficult to live with or critical, antagonistic, and alienated from parents and adults generally. These youngsters become strongly peer oriented, but at the same time they may experience difficulty in their peer relations as well. Such individual differences would be difficult to trace to their origin. Differences in congenital temperamental predispositions undoubtedly are often important factors. And, as Erikson (1968) suggests, experiences in early parent-child interaction, which contribute or do not contribute to a sense of trust in the environment, such as the care of a mothering person during early infancy who is "warmly dependable," may also be at the basis of a difficult identity crisis.

Adolescent Sensitivity to Deviations in Physical Traits

As was noted in young people, although they may want to be different to a degree-to express their individuality-also have a strong need not to deviate too much physically from the currently popular stereotype. Girls entering adolescence, as a rule, particularly if their inherent body type tends to be in the direction of endomorphifan, react quite negatively to a condition of overweight. Neither do they want to be too tall. Boys, on the other hand, are likely to be disturbed by an unusually short stature, They are especially sensitive to and disturbed by signs of sexual immaturity due to the late onset of pubescence.

This latter problem was noted particularly in a pair of brothers, ages 12 and 14, whose development was followed longitudinally. The older of the two entered the pubescent phase somewhat later than average, while his younger brother tended to be an early developer. As the older boy approached his fourteenth birthday he still maintained all the characteristics of a prepuberal child—a relatively short stature, a high-pitched voice, and no obvious signs of genital organ growth. At the same time his 12-year-old brother was rapidly exhibiting all the signs of puberty. For a period of several months this contrast in maturational status was a source of considerable anguish on the part of the older sibling. In an interview he expressed much concern as to whether he was "ever going to grow up." This disturbance was reflected in his relations with his family, particularly with his brother, and his age peers as well.

H. E. Jones (1949) in his studies of adolescence has pointed out that the effects of early or late maturing often affects boys and girls differently. Girls who mature earlier than average often experience it as a disadvantage in their relations with peers of both sexes. An early-maturing adolescent girl is much ahead of boys of her own age in size and in intellectual and social development. Normally adolescent boys and girls, even when they are of the same age, differ as much as a year or more in these areas of development. Such sex differences are often sources of some difficulty on the part of school personnel in dealing with mixed-sex groups of same-age youngsters.

In contrast with the early-maturing girl, the boy on an early-maturing schedule may experience some advantages. He will be larger and stronger than many of his age peers. Jones and Bayley (1950) found that early maturing boys were more attractive physically, less affected, and more relaxed when their behaviour was judged in comparison with non-early maturers.

Young people in early adolescence often are particularly sensitive to and concerned about relatively trivial deviations from the usual, or what they regard as desirable, in physical make-up. Such things as too slender or too plump legs, breasts that are underdeveloped or too large and pendulous in girls, and in boys, what they regard as poor muscle development can be matters of keen personal dissatisfaction. Bow legs and calves that are not sufficiently round are sometimes causes of anguish in either sex, especially during times when clothing styles call special attention to such irregularities.

Menstruation and its Psychological Impact

The menarche—the beginning of the menstrual cycle of a young girl—is the most reliable indication that she has reached puberty and is now capable of conceiving and bearing children. There is no strictly comparable criterion of sexual maturity in boys.

Menstruation is a natural phenomenon, a normal phase in the cyclic production and discharge of mature ova. In the past there have been many misconceptions and superstitions about this very important normal function. It has been referred to as a "monthly sickness" with which women are afflicted. Such views are not now widely accepted.

However, the fact that menstruation, which involves rather complicated bodily changes largely physiological in nature, begins at the onset of pubescence can have an important psychological impact, according to several studies . It is true that the process is frequently accompanied by unpleasant physical symptoms such as abdominal cramps, headache, backache, nausea, and fatigue. The attention of the pubescent girl is naturally drawn to this cyclical event in its association with puberty, and she may under certain circumstances regard it with resentment as a penalty for being a female. It is also true that even now in our culture menstruation is sometimes associated with considerable shame, embarrassment, secrecy, and dread. It is, however, accepted unemotionally and even eagerly by the majority of girls.

As was noted earlier, the attainment of sexual maturity, or the ability to reproduce, in boys has no single indicator criterion which compares with the menarche in girls for 'reliability. The full growth of genitalia and ejaculation of semen are perhaps the most definite indicators. These and other changes and events marking the developmental phase also have their potent psychological impacts. Voice change, clearly a result of structural maturation (a qualitative change in the vocal organs) is perhaps most embarrassing to the boy. The unpredictable "breaking" of the voice in social conversation is very disconcerting to some boys while to others it is nothing more than a slight annoyance.

Motor Functioning

We have just been concerned with the widespread and dramatic physical and morphological changes which characterize the approximately to-year period which we call adolescence. As we now center our attention more specifically upon the functional aspects of that development we shall note many equally interesting changes—changes in strength, speed, and smoothness of performance, as well as in the importance of motor abilities and skills in the adjustments of individual youngsters. We shall particularly note some important sex differences in the nature and direction of change in interests as well as actual abilities in areas of physical activity. Wide individual differences among both boys and girls in interest and ability to perform are seen in relation to school programs in physical education.

Changes in Motor Performance

The rapid growth of the body during the puberal phase, and the great amount of physical activity that continues to characterize adolescence, result in marked increases in physical strength. The youngster normally very soon becomes accustomed to his new and changing physical dimensions. Precision and smoothness in motor performance, as a rule, are soon developed. The adolescent's senses are keen and his reaction time is relatively short. Consequently, his speed of performance is generally high. Unfortunately, however, due to his lack of experience, his Judgment and discretion often do not match the level of his behavioural pace. He, of course, carries over his preadolescent interest in sports, and with his added size, strength, and speed of reaction his athletic skills generally are perfected. Thus we see a marked change from the so-called awkwardness of early pubescence.

These enhanced athletic abilities often play a significant role in social relations, particularly among boys. Comparable developments in girls, however, may play an equally important but more subtle role in the social development of adolescent girls. Clearly, nothing lends more to a boy's sense of importance and status during his high school years than recognized competence in athletics. As everyone knows, the football star is the envy of his male age peers and is often the idol of the female crowd.

Cognitive Development

From an earlier discussion it will be recalled that the course and progress of brain growth from birth to adulthood closely parallels that of intellectual development. It is only logical to assume that in the general speed up of body development, the puberal growth spurt described earlier, the brain would be involved. Students of cognitive development are in general agreement that a significant rise in quality of mental functioning also occurs during approximately the same period of time as the physical growth acceleration.

Influence of Piaget

We reviewed briefly Piaget's account of the development of mentality, or cognitive functioning. Cognitive development, as we have seen, begins with what is probably a complete lack of mental functioning at the time of birth and culminates in adult capacity to function with understanding, effectiveness, and facility in relation to the external environment. To reason, to solve problems through the study and manipulation of things, events, and relations not perceptually present is perhaps the most complicated and involved aspect of human development. It is an area in which much research has been done and is now underway.

Perhaps no other person has stimulated as much research or has directly contributed as much to the present level of understanding in this area as the Swiss psychologist Jean Piaget. His meticulous observations of child behaviour and his many "experiments" have led to the formulation of a complete and integrated theoretical account of cognitive development. Many other highly competent psychologists have made important contributions. Many of them have raised questions about Piaget's findings and theoretical statements, and many disagree with him at points. Piaget's account, nevertheless, continues in its general outline, to be widely accepted. We shall now briefly review the steps in the development of intellect during the period of adolescence as traced by Piaget and his co-workers.

Emergence of the Adult Level of Cognitive Functioning

According to Piagetian thinking, at about 11 or 12 years the youngster normally begins "freeing himself from the concrete" and orienting himself toward the non-present and the future (Piaget and Inhelder, 1969). At this level, however, he still thinks in terms of "concrete operations." Concrete operations provide a transition between schemata of action—the semiotic function-and the general logical structures, the "formal operations" of the mature intellect.

The child at this concrete operations level thinks in terms of mental representations. He is no longer limited to representing by "imitating" an object or even in the form of an overt act. The representation is "decentered" from action. His operations, that is, his thinking, are concrete, however, in the sense that they are tied to actual perceptual experience. He cannot yet deal mentally with propositions, with hypothetical situations not directly experienced perceptually. This level of thinking is exemplified in the simple "conservation" experiment in which all of the water in a glass container is poured, before the child's eyes, into another glass, which is taller and more slender. If the child is at the concrete operations level, he will say that, "it is the same water," "it is the same amount." He can represent mentally the attribute of a constant amount because he has perceived the dimensional changes in the water mass.

The transition from this stage to that of formal operations comes about gradually.

Formal Operations

The advance the young person makes in the quality level of his thinking as he frees himself from the "concrete"—from representations of things and of their perceived qualities and relations which he sees as true—and begins to reason correctly "about propositions he does not believe, or at least not yet; that is, propositions that he considers pure hypotheses" (Piaget and Inhelder, p. 132), is indeed remarkable. And, as was noted earlier, this step forward is quite contemporaneous with the generall pubescent acceleration of organic development.

The final "decentering" process in which one frees oneself for "propositional" thinking, for the consideration of all possible ways of solving a problem, requires time. It is not until about age 14 or 15, on the average, that the full level of formal operations is attained. The individual by that time can isolate the various elements of a problem and systematically explore a variety of approaches to its solution. The formal operations level of thinking thus makes possible the adoption of a generalized orientation to problem-solving. This may involve the controlled, critical evaluation of hypotheses, with final selection and testing-the thinking of the scientist.

Emotional Adjustments of Adolescence

According to Piagetian theory, this continuous yet sequential development of the cognitive functions is paralleled by an equally important emotional growth and increased capacity for social functioning. Along with the repeated process of "decentering" from earlier modes of behaviour toward more "mature" levels of cognitive functioning, this increase in capacity for higher level emotional and social functioning, can also be seen. For example, there is a gradual growth from complete egocentricity through varying degrees of capacity for sensitivity and empathy with respect to the feelings and welfare of others. There might also be growth in the capacity to defer gratification and to control and regulate the nature of one's emotional expressions with due regard to the situation and the presence of others.

Special note was made earlier of the psychological impact of growth and maturational changes of the puberal period. The changing pattern of endocrine glandular secretions, bringing about changes in the chemical content of the blood, result in new patterns of physiological functioning. These changes naturally affect directly the emotional experiences of the young individual. They give rise to the so-called sex drive and the whole related area of interest in and feelings toward members of the opposite sex.

Sex Differences—Changes with Development

During middle childhood there is usually little evidence of interest on the part of either boys or girls in the opposite sex. On the contrary, a kind of sex antagonism, real or pseudo, often builds up. Boys play with boys and organize their "no-girls-allowed" play clubs. Girls likewise prefer girls as companions. However, the situation begins to change in late preadolescence. Boys' interest in girls begins to develop. There are wide differences among boys as to the extent to which they manifest this interest, depending very largely upon their experiential backgrounds in family, school, and social groups. Some boys strongly deny and try to hide any interest they might have in girls. Other boys feel no need to hide their interest and they begin to enjoy friendships and association with girls. Prepuberal girls are more likely to frankly show their interest, especially in boys' activities. They more frequently participate, when allowed, in the rough-and-tumble games and stunts of the boys.

With puberal change, as already noted, come changes in emotionality in general, with a strong surge in interest in and need to associate with the opposite sex. At first this interest is likely to be quite general and nonspecific. There is simply pleasure in associating, talking together, bragging, "showing off," trying to impress one another. It is in such group associations, however, that the more intense mutual pair attachments develop. Pairing off and "going steady" tends to begin. However, by this time girls are one to two years ahead of boys in physical development. This results in a general age differential between boys and girls as they begin these more intimate relationships. Girls, who are also ahead of boys in social development, see the boys of their own age as crude and childish, and they turn their interest to older boys who are more nearly on their level developmentally. The "rejected" boys, the age peers of these girls, do not feel rejected. They are usually rather disdainful of the "put-on" airs of their former girl associates. A year or two later these boys begin to relate to girls somewhat younger than themselves.

Nature of Adolescent Sex Adjustment Problems

It is, of course, at the point when boys and girls begin to pair off and to establish the more intimate, one-to-one relationships that problems are likely to begin.

Depending on one's individual make-up and background of experience, any of a number of problems might arise when person-to-person intimacy begins. At first there may be simply the problem of just what does one do, how does one behave when alone with one of the other sex. Relationships established early, of course, are not all of the same nature. In some situations a rather serious problem of intelligently dealing with one's urgent "need" for sexual gratification develops. How this problem is dealt with, together with whether any inner conflict develops between what one does and one's internalized standards, depends upon factors in one's background.

As was suggested earlier, in the beginning of mixed-sex group associations the person-to-person relationships are generally of a very nonspecific nature. That is, the pleasure of associating together is largely based upon "differing social roles and complementary psychological natures." Boys and girls enjoy being together because they are of different sex and thus complement one another in the group situation. According to some students of adolescent behaviour, this same kind of interaction is sometimes carried over into the later, more intimate pair relationships (Staton, 1963). The couple find that they share many common interests and a mutual respect, and a closeness develops between them.

There are, of course, the more highly emotionally charged pair relationships. According to some observers (Grant, 1957; McKinney, 1960), these are of two general sorts, differing mainly in the quality of the predominant pattern of their emotional interactions and motives—the "amorous" or love attachment and the more specifically erotic or "genital" kind.

According to this view, heterosexual attraction may be based upon motives other than the specific desire for sexual intimacy. The amorous type of relationship presumably is one in which there is genuine and complete devotion of boy and girl to each other. There is sensitivity to each other's feelings, desires, and aspirations, and effort is made to facilitate each other m his or her personal fulfillment. There are mutual feelings of closeness and spontaneity in expressions of love and affection.

Not everybody, according to recent research, is capable of this kind of love experience. The findings suggest that such a temperament, that is, the capacity to give and to receive love, has its basis in sensory and affective experiences of early infancy. The work of a number of investigators indicates that the adequate meeting of the infant's need for appropriate stimulation and maternal care is the foundation for the development of affectional capacity.

The "genital" type of relationship, by contrast, is one in which the primary bond is the realization, or the promise, of mutual gratification of erotic love. There would be other expressions of regard for each other and other kinds of intimacy, but erotic gratification is the primary motive.

Sex Differences in Sexuality

So far our discussions of adolescent adjustment problems have proceeded with the 'implicit assumption that there is much similarity and mutuality between the sexes in manner and mode of coping with common problems. And indeed there is. But there are, nevertheless, certain problems and difficulties that are peculiar to the boy or the girl. Some of these, of course, are associated with a commonly shared area of difficulty; others are quite distinct and specific to either boys or girls.

With respect to the "sex drive," as such, for example, there are some rather significant sex differences. The main difference here seems to be not so much in the overall strength of the drive as in its immediacy and readiness of arousal. At any time and under almost any set of circumstances erotic feelings and interests may arise immediately in the boy's consciousness in response to visual, auditory, or other sensory stimuli. This aspect of the drive in boys is "imperious and biologically specific." The boy also seems subject in some degree to direct 'hormonal stimulation. Feelings thus aroused give rise to erotically charged thoughts and imagination. "He must confront (such arousal) directly, consciously, find within himself the means of obtaining discharge without excessive guilt, and means of control without crippling inhibitions"

In girls, on the other hand, the early "sex drive" is not specifically erotic in nature. The girl is likely to experience, instead, rather "disturbing longings, unknown stirring of feelings, feelings of tenderness and soft affection, a desire to love someone ... " (Staton, 1963, p. 351). Whereas boys readily become strongly aroused sexually through nothing more than thinking or conversation about girls or viewing female nude pictures, girls usually require much more direct stimulation of the erotogenic zones (lips, breasts, genitalia). "Their sexual feelings, too, are enjoyable for their own sake, constitute a pleasurable state of feeling to be maintained," rather than, as in boys, a feeling which immediately becomes erotic, a state of high tension which tends to mount, propels toward climax after which it rapidly subsides.

Individual Difference

There is, of course, wide individual variation in both boys and girls in the nature and strength of sexual motivation and expression. The evidence suggests, however, that the range of individual differences in this area is greater for girls than for boys. Boys more frequently and more closely follow the "typical" male pattern. Girls, on the other hand, show a wider spread of variation. A greater proportion of girls, at the older adolescent ages, in both "readiness" to function and in mode of sexual response tend to approach the pattern that is more characteristic of the male.

Sometimes from the dawning of physical signs of adolescence, and occasionally even before, a boy or girl will display unmistakable signs of erotic sexual desire and make physical approaches, seductive in the case of girls and aggressive in the case of boys, aimed at sexual activity.

Cultural Influences

Not only is the young person beset by compelling new physiologically based drives from within but he also finds himself on the threshold of a complex, inconsistent, and confusing adult world on the outside-a world with which he has had little involvement or concern heretofore. In his earlier, relatively carefree years of middle childhood he was not immediately faced with the problem of relating himself to and becoming a member of adult society. He was not yet concerned very seriously with his probable future status or his realm of functioning in it. His primary concern very largely had been his relations with his peers and his performance status among them. Because of this preoccupation with things that were of concern for him as a child, along with the fact that he received relatively little attention from adults, he finds himself on that threshold knowing little about the real day-to-day activities of adults, their concerns, their thinking. In most instances he has probably seen more of the sordid and the frivolous aspects of adult life, the recreational, pleasure-seeking activities. He is also likely to have observed and otherwise been made more aware of the competitiveness, the deceitfulness, and the lack of trustworthiness in society than of the truly social concerns, the honesty and integrity, that characterize most adults. In short he is not prepared for the adjustment problems he faces.

Self-identity Problem

Youth has always been a problem, which has concerned adult society. This concern, however, has been not so much with the inner perplexities and difficulties young persons themselves experience as with how to manage them, how to get them integrated with a minimum of disruption.

Of course, the young adolescent has his own problems. He almost suddenly finds himself with the bodily dimensions of the adults around him, and with his adult like mental capacity he senses what the expectations of society with respect to him will be. Yet at the same time his parents and others from whom he might expect to get most help probably continue to talk to him and generally to treat him as a child. Before he can do anything else he must understand these and many other contradictions. He begins a serious examination and reevaluation of himself, his present feelings, desires, and competencies in relation to those close to him and to society in general as he sees it. He begins the primary "developmental task" of adolescence.

This task is self-definition. Adolescence is the period during which a young person learns who he is, and what he really feels. It is the time during which he differentiates himself from his culture, though on the culture's terms. It is the age at which, by becoming a person in his own right, he becomes capable of deeply felt relationships to other individuals perceived clearly as such.

Thus in the later school years young people, beset with physiological revolution of their genital maturation and the uncertainty of the adult roles ahead, seem much concerned with faddish attempts at establishing an adolescent subculture with what looks like a final rather than a transitory or, in fact, initial identity formation. They are sometimes morbidly, often curiously, preoccupied with what they appear to be in the eyes of others as compared with what they feel they are, and with the question of how to connect the roles and tasks cultivated earlier with the ideal prototype of the day. In their search for a new sense of continuity and sameness, which must' now include sexual maturity, some adolescents have to come to grips again with crises of earlier years before they can install lasting idols and ideals as guardians of a final identity.

Erikson thus sees the achievement of a sense of identity not so much as the meeting of a sudden, full-blown "crisis" related only to the period of adolescence but rather as a long-term developmental process having its beginning in the earliest experiences of infancy, with "stages" in its epigenesis along the way. This crucial adolescent stage in the process normally results in a rather thorough revision of the self-concept.

As Erikson points out, an important factor determining the "satisfactory" working through of this problem is the adequacy with which one has been allowed to pass through each of the earlier critical periods of childhood and to achieve the related developmental tasks. It should be realized, too, that this universal problem, and responsibility of discovering one's true self and of working toward the fulfillment of one's potentials, really continues throughout life. However, a crucial step must be taken during the critical, transitional period of adolescence, but the process of self-discovery is generally not completed "once and for all" at adolescence.

An important facet of adolescent change, certain aspects of which we have already noted, is an upsurge in the so-called sex drive. One facet of normal erotic expression is the experience of "tenderness"—of loving and cherishing another individual as a person. As was noted earlier, this quality of tenderness seems to be more characteristic of girls than of boys. It will be recalled also that much emphasis has been placed upon the importance of early infancy as a time when the foundation is laid for the capacity to give as well as to receive affection and love. A wide range of differences is evident in individual level of this capacity. However, along with the adolescent advance in mental capacity, and associated with. the simultaneous upsurge in erotic functional capacity, is growth in the capacity to love. The adolescent, then, must deal with this new urge to erotic love as an attribute of himself in his task of coming to know himself.

Difficulties

Reference to the ego-identity problem suggests a single common crisis of adolescence. The problem is, of course, a common one, in the sense that certain of its broad aspects are essentially the same for all. But since every person at the beginning of adolescence, or at any other level of development, is an individual, different from every other, the particular identity "crisis" he faces is in many respects a highly personal one. It differs as a problem in its total pattern and in its details as seen by the young person himself. It is personal in the nature and the degree of inner conflict and travail it involves, and it is individual in the pattern of its resolution—in changes, which may ensue, in the direction of behaviour, overt, emotional, social.

Indeed, a goal toward which adolescent self-examination is implicitly or explicitly striving is the discovery and the expression of individuality. The young person, with his newly acquired ability to think critically about his own thinking—his feelings, desires, aspirations—strives to discover the various facets of himself and then to be himself through self-expression, all this in relation to the world of people to which he must relate and harmonize himself.

It is, of course, at this point of relating to society that youth encounters his greatest difficulty. He is by nature inclined to be idealistic. He wants love and trust in his relations with others. He looks for truly loving and valuing behaviour in the mature models about him. He looks for honesty, trustworthiness, genuineness, integrity, but too often and in too many places he sees quite the opposite in human interaction-in business, politics, government, education, and organized religion.

These discrepancies between his ideals and what he finds in society sometimes lead a youth to become critical and contentious. The erroneous assumption of a general "conflict of generations" at the time of adolescence has developed.

It is true, of course, that adolescent contentiousness sometimes takes the form of a rebellion against parental restrictions or an open conflict of ideas between the youngster and his parents. The parents, their ideas and motivational systems, sometimes become equated with those of adult society in general, and the conflict may lead to the youngster's complete rejection of the adult world. He turns to his peers for support, and like-minded peer groups attempt to establish subcultures of their own.

Young people with different temperamental natures and experiential backgrounds, however, react differently to essentially similar situations. Some may lead out in open protest as, for example, in "activist" movements among students. Others are inclined to react with hostility and destructiveness. This type of "conflict," of course, is one, which has stirred society to action. The rise of "juvenile delinquency" and crime has led not only to corrective and punitive measures against the young offenders but also to efforts to determine some of the causes of delinquency and to institute preventive measures such as providing better facilities for the constructive use of leisure time. Psychological research has thrown much light upon the various expressions of adolescent ego disturbance and the many factors underlying them.

Nature of the "Adolescent Crisis"

Our concern in this chapter thus far has been primarily with the developmental aspects of the period we call adolescence. We have noted at points certain essential relations between these various aspects of change the relation between physical structure and physiological functioning, for example, and the "psychological accompaniments" of physical and physiological change. All of these changes, taking place as an integrated process in a relatively short period of time, constitute a complex and often difficult problem-perhaps more aptly, a set of difficult problems-for the child and for those about him. There are, of course, the problems of psychological adjustment which beset the young person himself" problems of parents and others who relate to him, trying to understand more clearly his specific needs, and the broader societal problems-problems of education, problems of environmentally providing for healthy adjustment and preventing maladjustments.

The degree to which such problems can be intelligently coped with obviously depends upon the clarity with which the real nature of the "adolescent crisis," with its complex dynamics and its many variations, is comprehended.

The Traditional Conception

Many studies of adolescence, psychological, sociological, and anthropological, have been made through the years. Adolescence has been studied as a period of development, as a crisis period in the child's life and in his interpersonal and social relations, and also as a social problem confronted by each succeeding adult generation. Consequently, an overall "traditional" conception of the nature of adolescence has become quite thoroughly established. The following statement is an attempt to characterize that traditional conception.

As a result of physical growth changes, and changes in physiological functioning the child experiences a complex of psychological accompaniments—feelings, emotional stirrings, motivating "drives"—that are completely new and strange to him. He even feels strange to himself as a physical being. He is disturbed by the new quality of his emotions. He experiences an upsurge of erotic desires along with the resurgence of the old so-called Oedipus complex-his earlier erotic attachment to his mother and rivalry feelings toward his father. These feelings he finds inconsistent with the ideals and standards of right and wrong, which he had previously internalized, and so there is conflict. Anxieties, feelings of guilt, apprehension, resentment arise. He finds himself in a state of bewilderment, a state of "storm and stress."

His sense of growing up also makes him realize that he is approaching adulthood, and therefore he must soon assume that role and meet the expectations associated with it. He consequently becomes more concerned with the nature of the adult world, and with how he will be able to relate to it and function in it. But with his idealism he is likely to see little about the world to be pleased with. He is likely to become disenchanted with the adult generation (the "generation gap"). He begins to feel pressure to detach himself from his family physically and emotionally. He must discover "a means of escaping his dependent status in the family, and even more urgently the dimly recognized drives and feelings toward his parents. This is the psychological irritation which pushes the child from home, leading him to negotiate or battle with the parents for greater freedom."

All of these new psychological experiences combine to lead the youngster to a serious self-examination. He must try to make some sense of it all. He is faced with ego identity crisis—the problem of relating past to present, to future. In facing this problem he sees no source of help and support except age peers—others who presumably also are dealing with their identity problems. Hence the peer group takes on its crucial importance to him. The peer group and culture supplant the family as an anchor. It "provides a haven in which the delicate task of self-exploration and self-definition can be accomplished".

The reports of social workers and clinicians who daily work with individual problems of disturbed youngsters and their families, of course, differ in details, but in general terms they correspond rather closely to the above picture of the adolescent in crisis. Theoretical writers, generally drawing on clinical studies for their data, also describe the adolescent predicament, with certain individual and cultural variations, in the same general terms.

The general implication in the literature about adolescence is that this growing-up crisis is universal and not to be escaped—that all youth actually experience the "storm and stress" syndrome, disenchantment with the "establishment" and the need for detachment from family in their drive for autonomy. Although most writers suggest that the problem of the female adolescent may be different in detail from that of the male, the description generally is spelled out in terms of the male youth. What seems to be indicated here is the need for more "normative" data based upon samples of youth, which more fairly and adequately represent the general adolescent segment of the population. Clinical studies, without doubt, most accurately portray the nature of the problem or problems with which the clinician works, but clinicians do not work with individuals who are not embroiled in problems of maladjustment serious enough to require therapy. They deal with pathology, and their descriptions and generalizations may not apply at all to the non-pathological majority. The need is for more careful studies that are not subject to the "pathologist's error."

Study of "Normal" Youth

Unfortunately, not many extensive normative studies have been made of the psychological development and adjustments of young people during adolescence. The major longitudinal projects of the past half century, of course, have made important contributions, particularly in their portrayal of individual psychological development in relation to family and other cultural influences. But these studies of relatively small and rather highly "selected" groups may not be fairly applicable to any general population. Moreover, they often furnish relatively little data, which are directly comparable from case to case concerned specifically with the problems and development of adolescence. Consequently the definitive generalizations of these studies, valuable and enlightening as they are, may not aptly portray adolescence generally.

Some rather extensive surveys of the attitudes and adjustments of adolescent boys and girls have been conducted by the Survey Research Center of the University of Michigan (Douvan and Adelson, 1966). The data for these studies were obtained by interview involving some 1045 boys and 2005 girls from school grades 6 through 12. Careful sampling techniques were used in the selection of the subjects. The individual interviews were from one to four hours in duration, conducted at the schools. This means, of course, that the study was limited to students in school, which rather definitely "under-represents 'problem' children in the population at large".

The findings of this research, then, have reference specifically to ordinary non-pathological, school-attending youth of the late 1950's. The following paragraphs present a brief summary of the results and the conclusions based upon them.

Forward Thrust of Adolescence

The tendency in ordinary American youth, according to these studies, generally was not to turn their backs upon the prospect of a life in the world of adults—to become "alienated" and without a goal in relation to society as it exists. On the contrary, the adolescent "is both pushed and pulled toward a future" by forces within him. He does experience the organically aroused "psychic conflicts" associated with puberal change. The "regressive dangers" sensed in connection with these conflicts "assemble the power of the past to urge the child to leave family and childhood". There was, in other words, the urge toward the independence of adulthood. It was, however, a pull in the positive, forward direction rather than a denial, a rejection of something regarded as inherently evil to be repudiated. To the American adolescent, then, the future was by no means a remote or irrelevant prospect.

... It is crucial as it is absorbed, integrated and expressed in current activities and attitudes. In one form or another the future orientation appears again and again as a distinguishing feature of the youngsters who are making adequate adolescent adjustments.

Sex Differences

This forward thrust was found to be common to both boys and girls, but its style and focus differ markedly between the sexes. Boys were concerned with quite different kinds of future activity than were girls. Boys' choices were concrete and reality oriented. They thought in terms of preparation for specific work roles in relation to their own interests, tastes, and capabilities. Their choices, as a rule, were not colored by "dreams of glory." They thought in terms of work not much above the level of their own fathers occupations. The findings also suggested a relation between the clarity of vocational goals, along with realistic notions of what it takes to achieve those goals, and patterns of personal adjustment in their current lives.

Girls, on the other hand, according to this study were not strongly oriented toward jobs and vocational activities. Their focus was upon the interpersonal and social aspects of future life. Marriage was the common goal, with special concern in their thinking with the roles of wife and mother. However, they did have ideas about "instrumental acts" appropriately leading to the primary goals. Thus their choices of vocations were those which express their feminine interests and which were likely to place them in settings, which will bring them in contact with prospective husbands. Again, the results indicated a relation between clarity of goal aspiration and personal adjustment.

A clear concept of her adult femininity, of feminine goals and interpersonal skills, functions for the girl like the vocational concept for the boy. It bridges the worlds of adolescence and adulthood, brings the future concretely into current life, and allows the future to contribute meaning and organization to adolescent activities and interests. Girls who have relatively clear notions about the goals in adult femininity show a high degree of personal integration. Those girls who specifically reject a feminine future are troubled adolescents.

Presumably the implication here is that these girls are "disturbed" in the sense of a realization of, and rebellious concern about, the inequities in the currently prevailing sex roles in society.

The authors see the expression of these sex differences in their characteristic mobility aspirations. Those of boys typically were very realistically oriented. They were seen as "the concrete expression of a boy's faith in himself. The goal he chooses is realistic in light of his talent and opportunities, but is not overblown. And is cast in the phrasing of reality-what is the job like?" The girl's mobility aspirations were less tied to reality. Since their future mobility up or down depends upon the husband they get they "may as well dream big".

Another difference, which showed up between the sexes was in relation to the felt need to detach oneself from the family and to develop a measure of independence. In the first place, the researchers saw in general a less dramatic, and less conflictful process "than tradition and theory hold." They also noted a significant sex difference. The urge to be free and "to be one's own master is almost exclusively a masculine stirring". The girls of the study, up to the age of 18, showed no significant drive for independence.

Closely associated with the urge for independence was the issue of the importance of the peer group. A conclusion of this study was that in theory the importance of the peer group to the ordinary adolescent has been much exaggerated. There was also an interesting sex difference in allegiance to the group. The boys more frequently saw "the gang" as a loyal group who could give support to its members. Girls, they concluded, may use their peer relations in quite a different way. Two-person friendships were more common. The girl more frequently needed and sought the loyalty and support of a best friend. This person-to-person shared intimacy transcended the importance of the group in meeting needs of girls.

Identity Problem

With respect to the search for personal identity Douvan and Adelson (1966), from the results of their studies, characterized this problem also largely in terms of differences they noted between boys and girls. They conclude that "there is not one adolescent crisis, but two major and clearly distinct ones—the masculine and the feminine".

The identity problem is also phased differently for boys and girls in our culture, and the distinction again revolves around their different requirements for object love and for autonomy. We have noted that feminine identity forms more closely about capacity and practice in the personal arts, and we have seen in our findings evidence that the girl's ego integration co-varies with her interpersonal development. Masculine identity, in contrast, focuses about the capacity to handle and master nonsocial reality, to design and win for oneself an independent area of work which fits one's individual talents and tastes and permits achievement of at least some central personal goals. The boy's ego development at. adolescence already bears the mark of this formulation and reflects his progress· in mastering it. Identity is for the boy a matter of individuating internal basis for action and defending these against domination of others. For the girl it is a process of finding and defining the internal and individual through attachments to others.

The overall view of the common adolescent crisis presented by Douvan and Adelson, as they found it among ordinary school-attending American adolescents in the late 1950s, clearly differs in significant ways from the "traditional" view. In the latter view the youngster is described as becoming intensely concerned with ethical values, religious beliefs, and political ideologies. His need to detach himself from his family comes about largely because of a value crisis—a conflict between his idealism and the views of his elders, the "establishment." The interview material of these investigators supports the view that" American adolescents are, on the whole, not deeply involved in ideology, nor are they prepared to do much individual thinking on value issues of any generality" (p. 353), probably because to do so would endanger their relations in the community and complicate their personal problems of ego synthesis. These defenses, they concluded, tend to curtail experience and actually to hamper growth and self-differentiation. The traditional pattern of conflict and detachment "can be found in adolescence, but it is found in a bold, sometimes stubborn, and often unhappy minority". These bold ones are the ones who, according to Friedenberg

(1969), constitute the diminishing group of young people in our society who succeed in actually "identifying" and defining their true selves. They are Friedenberg's "vanishing adolescents."

Social Change and Adolescent Adjustment

Nearly 20 years have elapsed since the Michigan survey of adolescence was made. During this period of rapid social change much has happened. Times have changed. The general social environment in many respects is quite different now and is constantly changing. It is reasonable to expect, therefore, that in general the attitudes, aspirations, and common behaviour patterns of young people are also quite different from what they were 20 years ago.

In this connection it is important to return briefly to a general viewpoint which has been maintained throughout this text, namely, *(1)* that developmental change results from the joint influence of biological nature and the currently effective environment, and *(2)* that the relative weight of the environment in its influence upon change varies with the nature of the particular variable under consideration.

As to the adjustments and development characteristics of the adolescent period, investigators and authoritative writers have taken different positions on the question of biology versus environment in terms largely of their disciplinary biases. Cultural anthropologists and sociologists generally, for example, have been inclined to give much weight to the cultural factor in relation to the crisis—the "storm and stress" of adolescence assumed to be common in Western society. Margaret Mead (1928), for example, found little evidence of difficulty or conflict in the lives of adolescent girls in the relatively simple and conflict-free Samoan society. There was no implication, of course, that these girls did not undergo the biological changes of puberty, and experience the natural "psychological accompaniments" of these changes.

For Samoan children, however, there was no mystery about physical sex differences. Boys and girls grew up in close association. Sex interest and experimentation was taken for granted.

Although Samoan society was monogamous, both marriage and separation were easy and common. For the wife and mother, at the time of a mutually agreed—upon separation it was simply a matter of changing residence from one extended family or village to another.

Life in Samoa involved few social and occupational role choices, especially for girls. The adolescent girl soon became a wife and mother and lived securely in a large family. With so few alternatives and variations in life, the "identity problem" was simple and constituted no problem. Hence the cultural factor is indeed important. It can have much to do with the ease or difficulty with which the biological aspects of growing up are coped with.

In contrast with Samoan society of half a century ago, present-day Western society presents a kaleidoscope of constantly changing patterns of social and occupational alternatives which greatly complicate decision making for the adolescent. It is also understandable that such complications create strain and conflict within such long-established customs and institutions as marriage. Attitudes toward traditionally sanctioned relations between the sexes have changed and are changing, and a variety of alternatives to traditional marriage is currently being tried.

Such apparent breakup of the institution of marriage and family living without doubt is having its effect upon the personal adjustments of adolescents. Their patterns of thinking and their levels of sophistication are undoubtedly different than was the case 20 years ago. Their attitudes generally regarding relations between the sexes have been influenced. It is interesting to note, however, that in spite of these changes and their influence upon attitudes and behaviour in general the most recent evidence indicates that there is still a basic interest in, and need among young people for some form of stable marriage which is widespread and as strong as ever. In a summary statement D. H. Olson (1972) wrote:

> ...marriage still continues to be the most popular voluntary institution in our society with only three to four percent of the population never marrying at least once. And national statistics on marriage indicate that an increasing number of eligible individuals are getting married, rather than remaining single (Vital Statistics, 1970). For the third consecutive year, in 1970 there were over two million marriages in the country. The rate of marriage has also continued to increase so that the 1970 rate of 10.7 marriages per 1000 individuals was the highest annual rate since 1950. There has also been very little change in the median age of first marriage between 1950 and 1970, 22.8 to 23.2 years for males, and 20.3 to 20.8 years for females respectively (Population census, 1971).

As another indication of the effects of general attitudinal change, along with the increase in marriage rate, the rate of divorce has also changed. To our knowledge, no one has determined unambiguously the origin and development in the human being of this apparent need for the marriage relationship. It is evidently deep-seated and quite resistant to environmental change. In view of all the evidence it is probable that the reality oriented aspirations of adolescents of the 1970s are not greatly different than they were in 1950, in spite of radical social change in general.

●●

2

Psychology of Human Mind

Philosophical psychology, concerned with the epistemological problem of the nature of knowing mind in relationship to the world as known, contributed fundamental questions and explanatory constructs; sensory physiology and to a certain extent physics contributed experimental methods and a growing body of phenomenological facts.

In one version of this story that can be traced back at least to Ribot (1879), the epistemology of the 17th and 18th centuries culminated in the work of Kant, who denied the possibility that psychology could become an empirical science on two grounds. First, since psychological processes vary in only one dimension, time, they could not be described mathematically. Second, since psychological processes are internal and subjective, Kant also asserted that they could not be laid open to measurement. Herbart, so the tale goes, answered the first of Kant's objections by conceiving of mental entities as varying both in time and in intensity and showing that the change in intensity over time could be mathematically represented. Fechner then answered the second objection by developing psychophysical procedures that allowed the strength of a sensation to be scaled.

While there is undoubted truth in the received history, like all rationalizing reconstructions, it tends greatly to oversimplify what is an exceptionally complex story. Within the past 20 years, as primary resource materials have become more widely available and as larger numbers of historians have entered the arena, the received view has been amended many times. Within the context of this exhibit catalogue, it will not, of course, be possible to address this complexity. The reader who is interested, however, is referred to the Journal of the History of the Behavioural Sciences and to Bringmann & Tweney (1980), Danziger (1990), Rieber (1980), and Woodward & Ash (1982) among others.

Because so many psychologists are at least broadly familiar with the lines of Boring's story of the rise of experimental psychology, because the story has been so frequently retold in the man other textbook histories, and because it is a much more complex tale that it at first appears, this section and the two to follow will sketch only the barest outlines of the intellectual developments that led from Locke to Kant, from Bell to Muller, and from Fechner to Wundt. Psychologists who have not read Boring are strongly encouraged to do so. Despite its limitations, it is still the point of origin from which much of contemporary scholarship proceeds; and, perhaps even more importantly, it is the history of psychology that has become part and parcel of American psychology's view of itself.

Since we have already discussed Descartes and briefly touched on Leibniz, we can pass directly to the founder of both empiricism and associationism, John Locke (1632-1704). Locke was born in Wrington, Somerset, England reared in a liberal Puritan environment, and educated at Christ's Church, Oxford. His Essay Concerning Humane Understanding, dated 1690 but actually published in 1689, like much of the rest of 17th century philosophy, is a reaction to Descartes. Unlike Spinoza, who attacked the mind-body dichotomy metaphysically, Locke moved the discussion into the purely psychological realm of experience, contrasting inner sense (the mind's reflective experience of its own experience of things) with outer sense (the mind's experience of things).

Employing a very general notion of "idea" that incorporated a disparate set of entities among which modern psychologists would distinguish perceptions, mental images, and concepts, Locke concerned himself with both the certainty of our ideas experientially attained through reflection or the inner sense and the truth of our ideas insofar as they depend on the outer sense. After Locke, it would be possible to emphasize either the vivid character of the ideas transmitted by the outer sense or the intuitive certainty of the inner sense. The former view would lead to the sensationalism of Condillac, the later to the intuitional realism of Reid and the Scottish school of faculty psychology. In the 60 or more years intervening between Locke and Condillac, however, others, most notably George Berkeley and David Hartley, also made use of notions contained in Locke's Essay.

In the Essay on Humane Understanding, Locke had distinguished between primary and secondary qualities. Primary qualities such as solidity or extension are completely inseparable from the bodies in which they inhere and are simply perceived by the senses. Secondary qualities are the powers inherent in objects to produce sensations in the perceiver such as color, odor, or sound. The colors, odors, and sounds, however, do not themselves inhere in the objects. Berkeley's "immaterialism" was simply the notion of secondary qualities expanded to include primary qualities and taken out of objects and placed in God.

George Berkeley (1685-1753) was born at Kilkenny, Ireland and educated at Trinity College, Dublin. In 1709, he published his first book, Essay Towards a New Theory of Vision. Although Berkeley did not explicitly discuss his immaterialism in the New Theory, it was everywhere implicit in his views and combined with a proto-associational view of the importance of connections between ideas, it provided him with the basis for a theory of the perception of distance, which became a prototype for later associationist accounts. For Berkeley, distance is not immediately perceived by vision. Rather, when "the mind has, by constant experience, found the different sensations corresponding to the different dispositions of the eyes to be attended each with a different degree of distance in the object... (and) there has grown an habitual or customary connexion between those two sorts of ideas,... distance... is... the idea... immediately suggested to the understanding".

David Hartley (1705-1757) was born at Luddenden, Halifax, England and educated at Jesus College, Cambridge. In 1749, he published his two-volume Observations on Man. While the general principle of association was in use long before Hartley and the phrase, "the association of ideas," can be traced to the Appendix of the 4th edition of Locke's Essay, it is with Hartley, as Young (1970) tells us, that "the association psychology first assumed a definite form and a psychological character not wholly derived from epistemological questions. Hartley was the first to apply the association principle as a fundamental and exhaustive explanation of all experience and activity... Moreover he joined his psychological theory with postulates about how the nervous system functions.

Etienne Bonnot de Condillac (1715-1780) was born in Grenoble, educated in theology at Saint-Sulpice and at the Sorbonne, and ordained to the priesthood in 1740. Of the two sources of knowledge in Locke, sensations transmitted through the outer sense and reflection through the inner sense, Condillac focused exclusively on the former. His Traite des sensations, published in 1754, was designed to show that external impressions through the outer senses, taken by themselves, can account for all ideas and all mental operations. Using the famous example of a statue endowed with no other property than a single sense, smell, he attempted to derive attention, memory, judgment, imagination, the whole of mental" life. Condillac's views are, clearly, the most extreme form of the tabula rasa perspective. Like all tabula rasa views, no matter how powerful the correlative principle of association, Condillac's extreme sensationalism runs afoul of the obvious fact of variation (species differences, individual differences) in biological constitution.

In direct contrast to Condillac, Thomas Reid (1710-1796) chose to emphasize Locke's inner sense, building on the simple notion of reflection to develop an elaborate theory of the intuitions and faculties of the human mind given by its fundamental constitution. Reid was born near Aberdeen and educated at Marischal College. Initially influenced by Berkeley, his antipathy to the implicit assumptions in Hume's Treatise of Human Nature (1739) turned him away from both Berkeley and Hume and toward the reformation of philosophy.

In the Inquiry Reid articulated the basic intuitional postulate of the "common sense" philosophy on which the Scottish faculty psychology was to be built. Intuitions are native tendencies to mental action, aspects of the fundamental constitution of the human mind, which regulate the conscious experience of all human beings from birth. Because intuitions require the presentation of appropriate objects in order to be called forth in mental action, the Scottish philosophy is a realism. Intuitions do not project the mind into reality, they allow the mind access to it. Although intuitionalism is a nativism of psychological process, it is a methodological empiricism in that inquiry into the nature and existence of natively given principles of mind takes place by induction from observed facts in self-consciousness. It was this view, coupled with Reid's (1785-1788) later analysis of specific faculties that dominated 19th century academic American mental philosophy. It was also indirectly from Reid that Gall obtained the original list of 27 powers of the mind that guided his attempt to map the localization of function in the brain.

Immanuel Kant (1724-1804) was born, lived, and died at Konigsberg, in East Prussia. It is said that in the entire course of his life, he never traveled more than forty miles from the place of his birth. The suggestion from Ribot that 18th century philosophy culminated in the work of Kant was probably not an unreasonable one; although it might be an even fairer appraisal of Kant's influence to say that 19th and 20th century philosophy followed Kant much as the earlier philosophy had followed Descartes. Kant's indirect influence on scientific psychology was therefore enormous.

One such contribution, as we have already noted, was Kant's defining the prerequisites that would need to be met for psychology to become an empirical science. Another consisted of a bonafide psychological treatise, Anthropologie in pragmatischer Hinsicht, published in 1798. Long ignored, probably in part because of its pronounced sympathy for a soon to be discredited physiognomy, the Anthropologie is, nonetheless, a fascinating little book. Here Kant analyzes the nature of the cognitive powers, feelings of pleasure and displeasure, affects, passions, and character in the context of a denial of the possibility of an empirical science of conscious process. The Anthropologie went through two editions during Kant's lifetime and several later printings and helped to define the context within which not only Herbart and Fechner but phenomenologically oriented physiologists such as Purkyne, Weber, and Muller worked to establish the science of conscious phenomena that Kant was unable to envision.

The 19th Century: The Epistemology of the Nervous System

Boring (1950) has pointed out that between 1800 and 1850 discoveries in physiology helped lay the foundation for the eventual rise of experimental psychology. The events of particular interest to us are: *(a)* the first elaboration of a distinction between sensory and motor nerves; *(b)* the emergence of a sensory phenomenology of vision and of touch; and *(c)* the articulation of the doctrine of specific nerve energies, including the related view that the nervous system mediates between the mind and the world. While these discoveries were being made, two major developments in philosophical psychology were also occurring: the elaboration of secondary laws of association and the first attempt at a quantitative description of the parameters affecting the movement of ideas above and below a threshold.

The first of the relevant physiological discoveries, that of the distinction between sensory and motor nerves is credited to Charles Bell (1774-1842). Bell was born in Edinburgh and educated informally. Although he attended lectures at the University of Edinburgh, most of Bell's anatomical and surgical instruction was received from his older brother John, a noted physician. By the time Bell was in his twenties, he was already a well-respected surgeon and by 1799 he had been admitted to the Royal College of Surgeons in Edinburgh.

In 1806, he moved to London and five years later became affiliated with the Hunterian School of Anatomy. It was in that same year, 1811, that Bell printed one hundred copies of his 36 page Idea of a New Anatomy of the Brain for private circulation among his friends and colleagues.

In the New Anatomy, Bell employed anatomical evidence to support the assertion that the ventral roots of the spinal cord contain only motor and the dorsal roots only sensory fibers. In so doing, he overturned centuries of tradition in which it was implicitly assumed that nerve fibers were indiscriminate with respect to sensory or motor function and established the fundamental distinction between these two types of nervous processes. When, as we have already seen, this distinction was combined with a parallel sensorymotor associationism, it led in the hands of Bain and Spencer to the first properly psychophysiological psychology and, through

Jackson and Ferrier, to the establishment of the sensory-motor paradigm as the basis of functional localization in the cortex.

The first of the relevant philosophical advances was provided by, Thomas Brown (1778-1820). Brown was born at Kirkmabreck, Scotland and educated in philosophy and medicine at the University of Edinburgh where he took courses with Dugald Stewart, a disciple of Reid. In 1810, he was appointed to share the professorship of moral philosophy with Stewart and within a short time he had become renowned for the brilliance of his lectures. In 1820, after his premature death, these lectures were published in four volumes as Lectures on the Philosophy of the Human Mind. Their impact was immediate, undoubtedly because Brown managed to unite elements of two disparate traditions, the Scottish intuitionalism of Reid and the empiricism of Condillac. In so doing, he helped redirect both traditions.

Among a number of novel contributions, including an important critique of introspection based on Brown's belief in the absurdity of the idea that one and the same indivisible mind could be both the subject and the object of the same observation, Brown made two conceptual advances of fundamental importance in the history of experimental psychology. The first was to emphasize the "muscle sense". Before Bain, as we have earlier suggested, the associationists had neglected movement and action in favor of the analysis of sensation. Brown was the first philosopher in that tradition to move toward a more balanced sensory-motor view by including the sensory side of movement in his conceptualization of the problem of objective reference in perception.

Brown's second contribution involved his detailed elaboration of the secondary laws of association, which he termed "suggestion." Brown's formulation of these laws, which involved the relative duration, strength (liveliness), frequency, and recency of the original sensations as well as the reinforcement of one idea by others, provided later learning theorists with a basis for the attempt to explain not only the facts but the quantitative parameters of association.

During almost the same period, in Germany, another philosopher of mind, Johann Friedrich Herbart (1776-1841) was also concerning himself with quantitative relationships among ideas. Herbart was born in Oldenburg and studied at the University of Jena under Johann Gottlieb Fichte, with whom he found himself in some disagreement. Provoked by Fichte's ideas, Herbart decided to work toward his own systematic philosophy and upon completion of study at Jena, went to Gottingen where he took the doctorate in 1802. There he remained until 1809 when he moved to Konigsberg to assume the chair formerly occupied by Kant.

At Konigsberg, Herbart began work on his psychology, publishing his Lehrbuch in 1816 and Psychologie als Wissenschaft in 1824/1825. As is evident from this later title, Herbart believed that psychology could be both empirical (although he denied the possibility of experiment) and mathematical. Arguing that ideas ("presentations") are arrayed in time and vary in intensity, he attempted to create both a statics and a dynamics of mind and employed complex mathematical equations to describe an hypothesized system of principles of interaction among ideas.

Specifically, Herbart assumed that ideas of the same sort oppose one another while ideas of different sorts do not. Opposition progressively weakens the original idea in consciousness and, as a result, it eventually sinks below the threshold of awareness where it remains until the appearance of a similar idea in experience causes the original to rise at a speed proportional to the degree of similarity between the two ideas. Furthermore, as the original is pulled up by the new idea, similar ideas cling to it. Thus no idea can rise except to take its place in the unitary mass of ideas already present in consciousness.

In these views, Herbart took several giant strides along the path that the new scientific psychology would eventually follow toward a complex, carefully worked out, quantitative recognition of the critical distinction between ideas above and below the threshold of consciousness. As the received history suggests, he was a transitional figure between Kant and Fechner; but in his rejection of the possibility of experimental verification and his inability to link his philosophy of mind to the physiology of the brain, he travelled only part of the way toward the "new" psychology. Before psychology could be taken into the laboratory, it needed methods; and the primary source of the early methods lay not in the philosophy of mind, but in the work of physiologists such as Purkyne and Weber, who made fundamental contributions to the experimental phenomenology of sensation, and Muller, who elaborated the doctrine of specific nerve energies that systematized the epistemological role of the nervous system as intermediary between the mind and the world.

Jan Evangelista Purkyne (1787-1869) was born in Libochovice, in Northern Bohemia and received his first formal education at a Piarist monastery. After completing the novitiate, he spent a year in study at the Piarist Philosophical Institute. In 1807, under the influence of the writings of Fichte, he left the order and traveled to Prague. Two years of work at the University of Prague and an additional three years as a private tutor preceded his decision to return to the university to study medicine. In 1819, at the completion of his medical studies, he published his doctoral dissertation, Beitrage zur Kenntnis des Sehens in subjectiver Hinsicht. This led in 1823 to his appointment as Professor of Physiology at the University of Breslau. In that same year, he reprinted his dissertation as the first volume of Beobachtungen und Versuche zur Physiologie der Sinne. The second volume, which followed in 1825, was sub-titled Neue Beitriige zur Kenntnis des Sehens in subjectiver Hinsicht.

The two volumes of the Beobachtungen are among the great intellectual achievements of the period and constitute a major point of transition in the emergence of experimental psychology. With extraordinarily acute ability to observe phenomenological detail, Purkyne explored the psychological consequences in visual experience of a series of experimental manipulations of the conditions of stimulation, including application to the eyeball of pressure and electrical current, alteration in point of light exposure relative to the fovea, degree of eye movement, and variation in the intensity of light. While Purkyne is best known to psychologists for his classic descriptions of phenomena such as the change in apparent luminosity of colors in dim as opposed to bright daylight (the so-called "Purkyne effect"), it was the breadth and systematicity of his use of the experimental method to explore the parameters of sensory experience that helped lay the foundation for future laboratory work.

Ernst Heinrich Weber (1795-1878) was born in Wittenberg and educated at Leipzig, where he remained to serve as Professor of Anatomy from 1818 and of Physiology after 1840. In 1834, he published De pulsu, resorptione, auditu et tactu. In that portion of the work devoted to touch, Weber presented an extensive experimental exploration of the sensory phenomenology of tactile experience. Whereas Purkyne had shown the value of applying the experimental method to the phenomenology of sensation, Weber extended the approach beyond experimentation to quantification.

Coining the phrase, just noticeable difference (JND) to refer to the smallest perceptible difference between two sensations, Weber amassed data in support of the general principle that a JND in the intensity of a sensation is a function of the change in the magnitude of a stimulus by a constant factor of its original magnitude (ÆR/R). Although it has since been shown that there are significant limitations in the generality of this relationship not only across other sensory systems but even within touch itself, it would be hard to over-estimate the importance of Weber's discovery for the emerging science of psychology. In articulating the relationship which Fechner later termed "Weber's Law," Weber provided an existence proof for the possibility of establishing quantitative relationships between variations in physical and mental events. By linking these relationships to the nervous system, he helped, with Muller, to establish the epistemological function of the nervous system in mediating the relationship between mind and the physical environment.

Johannes Muller (1801-1858) was born in Coblenz and educated at the University of Bonn. He received his medical degree in 1822 and, after a year in Berlin, was habilitated as privatdozent at Bonn, where he rose eventually to the professoriate. In 1833, he left Bonn to assume the prestigious Chair of Anatomy and Physiology at the University of Berlin. His most important contributions to the history of experimental psychology were the personal influence that he exerted upon younger colleagues and students, including Hermann von Helmholtz, Ernst Brucke, Carl Ludwig, and Emil DuBois-Reymond, and the systematic form he gave to the doctrine of the specific energies of nerves in the Handbuch der Physiologie des Menschen fur Vorlesungen, published between 1834 and 1840.

Although Muller had enunciated the doctrine of specific nerve energies as early as 1826, his presentation in the Handbuch was more extensive and systematic. Fundamentally, the doctrine involved two cardinal principles. The first of these principles was that the mind is directly aware not of objects in the physical world but of states of the nervous system. The nervous system, in other words, serves as an intermediary between the world and the mind and thus imposes its own nature on mental processes. The second was that the qualities of the sensory nerves of which the mind receives knowledge in sensation are specific to the various senses, the nerve of vision being normally as insensible to sound as the nerve of audition is to light.

As Boring (1950) pointed out, there was nothing in this view that was completely original with Muller. Not only was much of the doctrine contained in the work of Charles Bell, the first of Muller's two most fundamental principles was implicit in Locke's idea of "secondary qualities" and the second-incorporated an idea concerning the senses that had long been generally accepted. What was important in Muller was his systematization of these principles in a handbook of physiology that served a generation of students as the standard reference on the subject and the legitimacy he lent the overall doctrine through the weight of his personal prestige.

After Muller, the two problems of mind and body, the relationship of mind to brain and nervous system and the relationship of mind to world were inextricably linked. Although Muller did not himself explore the implications of his doctrine for the possibility that the ultimate correlates of sensory qualities might lie in specialized centers of the cerebral cortex or develop a sensory psychophysics, his principle of specificity lay the groundwork for the eventual localization of cortical function and his view of the epistemological function of the nervous system helped define the context within which techniques for the quantitative measurement of the mind/ world relationship emerged in Fechner's psychophysics.

Mind, Brain, and the Experimental Psychology

It is in the work of Gustav Theodor Fechner (1801-1887) that we find the formal beginning of experimental psychology.

Before Fechner, as Boring (1950) tells us, there was only psychological physiology and philosophical psychology. It was Fechner "who performed with scientific rigor those first experiments which laid the foundations for the new psychology and still lie at the basis of its methodology".

Fechner was born in Gross-Sachen, Prussia. At the age of, 16 he enrolled in medicine at the University of Leipzig where he studied anatomy under Weber. No sooner had he received his medical degree, however, than his interest began to shift toward physics and mathematics. By 1824, he was lecturing in physics and in 1834, with over 40 publications to his credit, including an important paper on the measurement of direct current, he was appointed Professor of Physics at Leipzig.

Fechner's psychological interests began to manifest themselves toward the end of the 1830s in papers on the perception of complementary and subjective colors. In 1840, the year in which an article on subjective afterimages appeared, Fechner suffered a nervous collapse. Exacerbated by a painful injury to the eyes sustained while gazing at the sun during his research, Fechner's ailment manifested itself in temporary blindness and prostration. He resigned his position at Leipzig and went into a lengthy period of virtual seclusion during which his interests turned increasingly toward metaphysics. In 1848, the year of his return to the University as Professor of Philosophy, he completed Nanna, oder Uber das Seelenleben der Pflanzen, a metaphysical treatise that contains his first explicit, philosophical treatment of the problem of the relationship of mind to body.

In Nanna, and in the more important Zend-Avesta (1851), Fechner sketched out a dual-aspect, monistic, pan-psychical mind/body view. In a famous metaphor, later adopted by Lewes, Fechner likened the universe, which is at one and the same time both active consciousness and inert matter, to a curve that can be regarded from one point of view as convex and from another as concave yet still retains its essential integrity. In line with this approach to mind/body, Fechner laid out a future program for psychophysics—to demonstrate the unity of mind and body empirically by relating increase in bodily energy to corresponding increase in mental intensity.

Between 1851 and 1860, Fechner worked out the rationale for measuring sensation indirectly in terms of the unit of just noticeable difference between two sensations, developed his three basic psychophysical methods (just noticeable differences, right and wrong cases, and average error) and carried out the classical experiments on tactual and visual distance, visual brightness, and lifted weights that formed a large part of the first of the two volumes of the Elemente der Psychophysik. Fechner's aim in the Elemente was to establish an exact science of the functional relationship between physical and mental phenomena. Distinguishing between inner (the relation between sensation and nerve excitation) and outer (the relation between sensation and physical stimulation) psychophysics, Fechner formulated his famous principle that the intensity of a sensation increases as the log of the stimulus (S = k log R) to characterize outer psychophysical relations. In doing so, he believed that he had arrived at a way of demonstrating a fundamental philosophical truth: mind and matter are simply different ways of conceiving of one and the same reality.

While the philosophical message of the Elemente was largely ignored, its methodological and empirical contributions were not. Fechner may have set out to counter materialist metaphysics; but he was a well-trained, systematic experimentalist and a competent mathematician and the impact of his work on scientists such as Helmholtz, Ernst Mach, A.W. Volkmann, Delboeuf, and others was scientific rather than metaphysical. By combining methodological innovation in measurement with careful experimentation, Fechner moved beyond Herbart to answer Kant's second objection regarding the possibility of scientific psychology. Mental events could, Fechner showed, not only be measured, but measured in terms of their relationship to physical events.

As Fechner was putting the finishing touches on the Elemente, a young physiologist, Wilhelm Wundt (1832-1920), was settling into a position as assistant to Helmholtz, who had come to Heidelberg from Bonn to direct the Physiological Institute. Wundt was born at Neckarau, in the vicinity of Mannheim and received his early education at the hands of a private tutor and at the Bruchsal Gymnasium. At age 19, he set off to study medicine at Tubingen, where his uncle, Friedrich Arnold, held the Chair in Anatomy and Physiology. During his first summer semester, he worked intensively on the study of cerebral anatomy under Arnold's guidance and by the end of the summer he had decided to make physiology his career. When his uncle moved to Heidelberg to direct the Institute of Anatomy, Wundt followed, completing his medical studies in 1855. After a year of hospital work and a journey to Berlin for a semester of study under Muller and Du Bois-Reymond, Wundt returned to Heidelberg in 1857 as Dozent in Physiology, becoming assistant to Helmholtz in the following year.

During this period, Wundt seems to have availed himself but little of his contact with Helmholtz. Carrying out much of his experimental work in his own home and on his own time, Wundt began the study of sense perception that led to a series of publications collected, in 1862, as his Beitrage zur Theorie der Sinneswahrnehmung. The Beitrage consisted of six previously published articles on sense perception preceded by a methodological introduction. In these articles, Wundt provided the basics of a psychological theory of the perception of space (including some discussion of the need for unconscious inference, apparently arrived at in independence of Helmholtz), reviewed the history of theories of vision, analyzed the psychological function of sensations arising from visual accommodation and eye movement, presented the results of experiments on binocular contrast effects and stereoscopic fusion, and argued, contra Herbart, that the content of consciousness at a given instant always consists of a single, unconsciously integrated, percept.

Although the body of the Beitrage is important in its own right for exemplifying the direction that Wundt' work was taking, it is his introduction on method, written specifically for the Beitrage, which marked the emergence of Wundt's plan for an experimental psychology. Rejecting a metaphysical foundation for psychology, Wundt argued for the need to transcend the limitations of the direct study of consciousness through the use of genetic, comparative, statistical, historical, and, particularly, experimental methods. Only in this way, he, suggested, would it be possible to come to a needed understanding of conscious phenomena as "complex products of the unconscious mind" (p. xvi). As the young Wundt was engaged in thinking through the prerequisites of an experimental psychology, Helmholtz, his immediate superior, the Director of his Institute, was in many ways already engaged in carrying out such a programme.

Hermann Ludwig Ferdinand von Helmholtz (1821-1894) was born in Potsdam and educated at the Potsdam Gymnasium and at the Friedrich Wilhelm Medical Institute in Berlin. In Berlin, he came under the influence of Muller and in 1842, at 21 years of age, he graduated with a degree in medicine and entered the service as a Prussian Army physician. In reaction to Muller's vitalism, which he rejected, Helmholtz became interested in clarifying the physiological basis of animal heat, a phenomenon that was sometimes used to help justify vitalism. This led in 1847 to a famous paper on the conservation of energy, which in turn brought Helmholtz the offer of a Professorship of Physiology at Konigsberg, where he remained from 1848 to 1855. In 1855, he moved to Bonn and from Bonn, in 1858, to Heidelberg to serve as Director of the Institute of Physiology.

It was during the Bonn and Heidelberg periods that Helmholtz made his most fundamental contributions to the newly emerging experimental psychology. From 1856 to 1866, the Handbuch der physiologischen Optik appeared in parts that were gathered into a single volume in 1867. In 1863, while the Optik was still appearing, Helmholtz published Die Lehre von den Tonempfindungen. While we will focus on the Optik here, these two works taken together defined the problematic for the experimental psychology of visual and auditory perception for decades to follow.

In the Optik, Helmholtz extended Muller's doctrine of the specific energies of nerves to offer a comprehensive theory of color vision and a famous unconscious inference theory of perception. In the theory of color vision, Helmholtz reasoned that just as the differences between sensations of sound and light reflect the specific qualities of auditory and visual nerves, sensations of color may depend on different kinds of nerves within the visual system. Since the laws of color mixture suggest that virtually all hues can be obtained by, various combinations of three primary colors, it seemed to Helmholtz that the perceived hue, brightness, and saturation of color must be derived from varying activity in three primary kinds of nerve fibers in the eye.

In his theory of perception, Helmholtz started from the recognition that Muller's doctrine of specific nerve energies implied the fact that sensations do not provide direct access to objects and events but only serve the mind as signs of reality. Perception, on this view, requires an active, unconscious, automatic, logical process on the part of the perceiver, which utilizes the information provided by sensation to infer the properties of external objects and events.

In an earlier period, Helmholtz had also made another major contribution to physiology. Stimulating nerves at various distances from a muscle and measuring the time it took for muscular contraction, he estimated the rate of travel of the nervous impulse, and in the process incidentally introduced the technique of reaction time into physiology. Between 1865 and 1868, another great physiologist, Franciscus Cornelis Donders (1818-1889) assimilated the reaction-time procedure to psychology, employing it to study the time taken up by mental operations.

Donders was born in the town of Tilburg, in the Netherlands, and entered the University of Utrecht as a medical student at the age of 17. Upon receipt of the degree, he joined the military as a surgeon and, at the age of 24, was invited to teach at the Military Medical School at Utrecht. Five years later Donders was offered a position as extraordinarius at the University of Utrecht, which he accepted, remaining there for the remainder of his career.

In 1865, Donders published a preliminary communication in which he reported work carried out with a student, Johan Jacob de Jaager, and summarized more fully in de Jaager's doctoral dissertation, De physiologische tijd bij psychische processen (1865). Reasoning that reaction time was additive, Donders separately assessed the time taken to respond to a stimulus under conditions of choice and simple non-choice. Subtracting simple from choice reaction-time, Donders computed the interval taken by the decision process. In 1868, in a classic paper appearing in German, "Die schnelligkeit psychischer Processe", Donders provided the definitive report of the results of this work and its extension to discrimination times.

As Donders investigated reaction-time, Wundt, still at Heidelberg, began to work toward the conception of physiological psychology that was to serve as the basis for his systematic approach to experimentation. In 1867, in a new quarterly journal of psychiatry founded by Max Leidesdorf and Theodor Meynert, Wundt published an invited article, "Neuere Leistungen auf dem Gebiete der physiologischen Psychologie." Under the banner of physiological psychology, he reviewed recent literature on visual space perception and the measurement of the time taken by mental operations. As an outgrowth of this review, Wundt offered a series of lectures on physiological psychology in the Winter of 1867/1868. These lectures he repeated only once again, in 1872/1873, as he was preparing the text that Boring (1950), steeped as he was in the Wundt-Titchener tradition, called "the most important book in the history of modern psychology".

Issued in two parts, in 1873 and 1874, the Grundzuge der physiologischen Psychologie was the first comprehensive handbook of modern experimental psychology. It was, as Boring tells us, "on the one hand, the concrete result of Wundt's intellectual development at Heidelberg and the symbol of his metamorphosis from physiologist to psychologist, and, on the other hand... the beginning of the new 'independent' science". Although the theories elaborated in the Grundzuge changed over the five major revisions during which it grew from one to three volumes, the essential structure of Wundt's system, "his great argument for an experimental psychology", had been reasonably well worked out by 1874.

In that year, Wundt accepted a call to the University of Zurich, where he remained only a year, moving in 1875 to Leipzig to assume the chair in philosophy. Although Boring (1950) claimed that upon his arrival in Leipzig Wundt was allocated space for experimental demonstrations adjunctive to his lectures, there is no evidence to that effect (Bringmann et al., 1980). Indeed, it would appear that from 1875 to 1879, Wundt devoted himself largely to the duties entailed in his new teaching position.

On the 24th of March, 1879, however, Wundt submitted a petition to the Royal Saxon Ministry of Education in which he formally requested a regular financial allocation for the establishment and support of a collection of psychophysical apparatus. Although his request was denied, Wundt seems as early as the Winter of 1879/1880 to have nonetheless allowed two students, G. Stanley Hall and Max Friedrich, "to occupy themselves with research investigations". This research took place in a small classroom in the Konvict Building that had earlier been assigned to Wundt for use as a storage area. Humble though it may have been, this small space constituted the first laboratory in the world devoted to original psychological research.

●●

3

Human Behaviour and Emotions

Emotionality as an area of functioning is complex and difficult to investigate. Important as it is in our lives, when viewed objectively it is a behavioural area which man has in common with subhuman species. Furthermore, although man's cognitive processes are importantly involved, his affective experiences and his emotional behaviour are largely mediated by lower centers of the brain. These centers and the other bodily structures involved in emotional functioning are intact and in a state of readiness at birth. The human being is born well equipped for experiencing feelings of pleasure and displeasure and for displaying emotional behaviour. In that particular respect man is more like the subhuman species. In that particular respect also maturational changes between birth and adulthood are relatively minor. Hence, even though emotionality is complex and multifaceted, involving as it does the whole organism, it is the most "primitive" major area of human functioning.

Aspects of Emotionality

Emotionality has been defined in various ways and has been seen from different theoretical points of view by students of human behaviour. One of these approaches, objective in nature, is to define an emotion in terms of the external situation—the perceptual field that gives rise to it. Thus, those emotional reactions, which result from situations that are judged objectively to be dangerous would be labeled fear or terror. Likewise, reactions to confining or frustrating situations would be anger or rage, and so on. This was the approach used by early investigators of emotions in infants and children. Watson (1924), for example, in his efforts to account for all infant emotions, defined the three emotions fear, anger, and love in terms of the relatively few stimuli to which he found infants responsive. The infants response to a sudden loud noise or a sudden loss of support he labeled fear, their reactions to restraint he labeled anger, and responses to being rocked, to stroking, fondling, and the like, were love.

A second objective approach to the study of emotions, closely related to the first, is in terms of the observable reactions, or measurable bodily changes, that constitute emotional behaviour. Emotional behaviour from this viewpoint is both external and internal. External behaviour, of course, is quite readily observed. When the behaviour is destructive or assaultive in nature it is labeled anger or rage. When it appears as less violent it is affection or love or lust or even elation. When behaviour appears as retreat or flight it is fear. Quite obviously, the mere observation of external behaviour without more information, including a knowledge of the eliciting stimuli, is not a sufficient basis for an adequate classification of emotion.

The internal responses involved in emotion are very complex and, of course, cannot be observed directly. Their effects, however can be measured by the use of instruments. The study of the internal, or visceral, changes involved in emotion has contributed much to our understanding. Complicated sequences and patterns of change involving the vascular system, the endocrine system, and the involuntary muscular system are extremely important aspects of the complex we call emotion. These physiological reactions are controlled and regulated by the autonomic division of the nervous system. In a very general way the "approach" and "avoidance" aspects of behaviour are mediated by the parasympathetic and the sympathetic segments of the autonomic division, respectively. The internal aspects of emotional behaviour, however, are always widespread and diffuse, and so far only very rough correlations with specific nervous structures have been established.

A third approach to the problem of the emotions is to regard emotion as a conscious experience. In spite of efforts on the part of some students of human behaviour to avoid a consideration of the conscious aspect of emotions, no "normal" human being—including the objective psychologist himself—can doubt the reality of the consciously experienced emotion. When we act angry in response to an anger-provoking stimulus we also feel angry.

A Response of the Total Person

In view of these three approaches to the study of emotional behaviour, an important fact stands out: Emotionality is an extremely complex area of human functioning, which involves the whole organism. The stimulus aspect cannot be neglected. Cognition is always involved. There is no emotion without either the actual perception of the provoking situation or the anticipation of it in terms of images. Normally, a situation does not provoke genuine fear or anger or jealousy or joy or elation unless it is first perceived or thought of as something about which one must be fearful or angry or jealous or joyful or elated. Even in the case of irrational anxiety or worry, image's or other symbols representing past cognitive experiences or imagined ones are being centrally manipulated.

The external behavioural aspects, the so-called expressive movements of emotion, are equally essential to the total complex. In full-fledged anger, for example, although actual destructive aggression may be held in check, there are, nevertheless, implicit if not overt patterns of muscular contractions and gross bodily movements that express hostility and tendency to aggression. The kinesthetic stimuli produced by these expressive behaviours give rise to sensory experiences that are important ingredients in the total experiential aspect of the complex.

Simultaneously with overt behaviour come the internal responses, which are mediated by the autonomic division of the nervous system in response to cognitive awareness of the situation. The whole pattern of changes is "designed" to prepare the organism physically for biologically appropriate activity. In the case of fear or anger, the joint action of the autonomic nervous system and the endocrine glandular system supplies the blood with additional energy-giving blood sugar and produces changes in the pattern of blood-vessel tension so as to shunt that energy-laden blood into the skeletal muscles for greater strength in meeting the "emergency." Internal organs, like skeletal muscles, are equipped with sense receptors that are stimulated by the internal activity, thus giving rise to patterns of sensations, which add to the "unanalyzed mass" that constitutes the emotional experience.

A realization of the importance of these sensory components arising from response-produced stimuli was what led William James, in 1890, to make his famous paradoxical statement that "we do not tremble because we are afraid but we are afraid because we tremble." According to his theory, which was also stated independently by the Danish physiologist Carl G. Lange, an emotion is the way the body feels when in a disturbed internal state and while making various overt expressive movements. The massive, unanalyzed experience arising from the bodily processes and the expressive movements is the emotion, according to James.

The important point here is that an adequate consideration of an area of human behaviour as complex as the emotions must deal with overt behavioural, as well as psychological, aspects. We would agree essentially Ausubel (1958):

> Emotion may be defined as a heightened state of subjective experience accompanied by skeletal-motor and autonomic-humoral responses, and by a selectively generalized state of lowered response thresholds...
>
> The following sequential steps are involved in the instigation of an emotional response: *(a)* interpretive phase—perception or anticipation of an event that is interpreted as threatening or enhancing an ego-involved, goal, value, or attribute of self; *(b)* preparatory reactive phase consisting of a selectively generalized lowering of the particular response thresholds implicated in a given emotion; *(c)* consumatory reactive phase with subjective, autonomic-humoral, and skeletal-motor components; and *(d)* a reflective reactive phase involving subjective awareness of the drive state and of visceral and skeletal responses.

Positive and Negative Emotions

As is obvious and common in everyone's experience, there are two general directional types, or categories, of emotional behaviour and experience. These types are so different qualitatively that they seem, subjectively, and appear objectively, to have nothing in common. "Pleasant," "positive," and "approach" apply to the one directional type, while "unpleasant," "negative," "escape," "injure," and "destroy" apply to the other. Not only are they logical and subjective opposites but the two types of experiences also involve mutually antagonistic autonomic functioning, thus mediating overall patterns of visceral change that are quite different. Yet the common term "emotion" applies equally well and without conflict to both categories of experience. Both types are heightened states of subjective experience. Both are extreme deviations from the comfortable, normal internal state of affairs, both being, in a sense, "emergency" reactions.

As we shall see in other connections later, the relative dominance of one or the other of these two emotional types has important implications in the child's personal development. The prevailing nature of the young child's environment will have much to, do with determining whether the social aspects of his life become attached to the positive or the negative categories of emotion.

Study of Emotional Development

Emotional experiences and outward emotional behaviour are matters of universal, firsthand acquaintance. Daily we all experience a variety of emotions in ourselves, and we are constantly meeting the problem of dealing with them in others. Underdeveloped emotional capacity, uncontrolled emotional behaviour, and pathological expressions of emotion are major areas of concern among those who must deal with human problems at any level, whether individual, group, national, or international; hence the need for greater understanding of the nature and development of emotionality.

Controlled research in this area is extremely difficult. Behaviour under emotion-arousing conditions in natural settings can be observed directly and descriptions of that behaviour can be recorded. But affectivity itself—what the individual is actually experiencing under those conditions—obviously cannot be observed objectively. Affectivity must be inferred in terms of the situation and the observed external behaviour associated with the situation. We include in the behaviour that can be observed the individual's subjective report of his feelings.

Our greatest insights into the nature of emotionality have come from the clinical approach. In the clinical setting, real emotional problems of individuals with widely varying backgrounds and experiences are studied. The clinician must, in each case, understand and deal with a total, unique emotional situation. Clinical studies have brought to focus the facts of individuality in human behaviour and the great diversity of environmental influences that affect individual emotional development. They clearly portray, for example, the fact that each child has his own inner strivings and frustrations, his own "self image," his attitude toward himself in relation to others. The evidence suggests that nothing gives the child more help and support in resolving his personal problems, and thus facilitating his emotional growth, than the feeling that he is really understood, that someone important to him really knows how he feels.

In our longitudinal studies of individual children our aim is to approximate as nearly as possible a "clinical" understanding of the particular emotional behaviour patterns that characterize each child. The aim is to arrive at an estimate of his level of emotional development at specific points in time and thus to trace that development through time.

Congenital Affectivity

The first breath of the newborn infant appears to be an outward expression of an emotional disturbance—a "protest" against intolerable conditions. The newborn apparently keeps on breathing because breathing promotes a satisfying state of affairs. It eliminates discomfort.

During the neonatal period the baby normally shows signs of emotional excitement only when conditions, external and internal, are such as to cause him bodily discomfort or sensuous pleasure. When he feels pain from some physical disorder or external condition, when he is hungry, or when he. is wet or cold or otherwise uncomfortable, he "reports" immediately with signs of distress. No condition outside his own body which does not impinge directly upon his sense organs is of his slightest concern. He is completely egocentric in that sense; he cannot be otherwise. When he senses discomfort he reacts immediately. He "demands" immediate relief from discomfort and immediate gratification of his needs with loud and insistent crying. He does not have to learn these patterns; they are built-in.

Babies, of course, are not all alike. They vary widely with respect to a number of temperamental tendencies. Some are by nature highly active, others are unusually quiet; some of them almost from the beginning seem to be relatively undisturbed and calm in their new conditions of existence, others appear to be displeased and uncomfortable. Babies also vary widely in their reactivity to strong stimuli: The "excitement" pattern (reactions of startle, shock, and withdrawal), with its wide individual variation, is common in infants, particularly during the first month of life before the development of the inhibiting function of the brain.

Rather marked congenital differences in temperament were observed in our subjects Paul and Sally. Paul was relatively quiet, with an observant and "thoughtful" appearance, while Sally, from the beginning, was an active child, inclined toward "unevenness of disposition." She was more reactive to restrictions, with ready tears when her movements were hampered. Both babies were alert when awake and quite reactive to stimulation.

Congenital temperamental differences 'presumably constitute the beginnings of individual differences in later emotionality.

Role of Learning in Early Emotional Development

Emotional learning, however, gets underway very soon. The baby's emotional expressions begin to vary with the situation, the particular source or locus of discomfort, and he begins to show differentiated expressive signs of pleasure from gratification as well as distress from pain or discomfort.

As development continues with experience, he becomes responsive emotionally to a wider and wider range of stimuli. Thus the direction of the child's emotional development and the characteristic quality of his emotional expressions are influenced by his environment, and particularly by the kind of care and handling he receives. Once again, development is a product of organism-environment interaction.

The baby's responsiveness to other human beings soon becomes pronounced, and, although evidence is not always clear, the consensus is that the quality of the attention and care he receives during this early period has much to do with the development of his later capacity to accept love and affection from others or to feel love for others. The evidence for this facet of emotional development comes largely from clinical observations. The development of the capacity for love is discussed in greater detail in a later section of this chapter.

Development Sequences

For an accurate assessment of developmental status in emotionality, however, just as in other functional areas, some sort of scale or sequence of stages is necessary. As is true with other generalized qualities of behaviour, no completely satisfactory series of indicators has been established, although several formulations have been proposed. Probably the best known of these is the series of stages postulated by Freud.

The Freudian Stages

The series of stages in the psychosexual development of the individual adopted by many psychologists as the basic framework of personality development are the oral, the anal, the phallic, the latency, and the genital periods. Certain of these stages were briefly described in the discussion of the vital functions. We shall sketch this series of stages again specifically as an example of a sequence in the development of emotionality.

The oral period covers early infancy, when taking food through nursing at the breast and being made comfortable by this process are the totality of life. Much of the baby's experiences of release from tension and gratification come from or are associated with the oral activity of food taking. When gratification through nursing is not forthcoming, when tension has mounted, feelings of frustration are experienced; anxiety and the sense of insecurity arise. With gratification come pleasure and a sense of security. This period, during which emotional experience—either pleasurable or otherwise centers around infant oral activity, lasts to about the end of the first year, thus corresponding to the period of greatest dependency.

The generalized pleasure aspect of infant oral activity, quite apart from the mere relief from hunger, is emphasized by the psychoanalytically oriented students of child behaviour.

Throughout this early period the child is rapidly learning to do things for himself. He begins now to explore more widely as he masters the ability to move about on his own power. His abilities to manipulate with his hands the objects about him appear now to be replacing the pleasure he has been deriving from sucking his fingers. Furthermore, as he becomes more and more able to "get into things," restrictions are placed upon him. He begins to experience new demands and pressures. The attitudes of those around him regarding order and cleanliness begin to be felt. The job of toilet training begins in earnest. Since the eliminative functions pertain to and directly involve his own body and its sensitivities, they are "closer" to him than are any other activities. They involve tension, release from tension, frustrations, and gratification. The child's preoccupation is thus shifted from oral activity to the anal region. According to the Freudian account, the child has now reached the anal stage.

Certainly, this period in the child's life is extremely important from the standpoint of determining developmental trends, and it can be a very difficult period for both the child and his parents. By the time the child has reached the age of 3 years oral pleasures, as a rule, are no longer predominant in his life. He is also well along in achieving bowel and bladder control. Concentration upon anal activities has also lessened as his manipulative and locomotor abilities and his interests in things outside himself gain precedence. His curiosity about and interest in people also are expanding. He begins to observe people, particularly their behaviour toward one another. In the typical family situation, although he takes it for granted, he is aware of and enjoys the special kind of intimacy and the love his parents show for him and for each other; in this close relationship his love for his parents deepens. He identifies with his parents. The little boy wants to be like Daddy.

According to Freudian theory, however, a change begins to take place in the child's feelings toward his mother. He becomes especially strongly attached to her, and his attachment is erotic in nature. He is now entering the phallic phase of psychosexual development. Very strong "object cathexis" develops toward his mother. In his desire to have her to himself his father becomes his great obstacle. Despite his great love for his father he wishes his elimination. He is in the midst of the Oedipus complex.

This naturally becomes a very conflictful situation. He experiences "bad" wishes. He desires incest with his mother and the elimination, even death, of his father. According to theory, he may also experience deep anxiety. His father is powerful. As his rival, his father, he fears, might retaliate and rob him of his phallus. Castration anxiety may thus develop.

In his need to escape from the stressful predicament certain "mechanisms" may come to the child's aid. Repression of his shameful, guilt-arousing wishes, along with the childish associations and memories connected with them, takes place. At the same time, he may gain some degree of satisfaction of his need for his mother by strongly identifying with his father, the one who possesses her.

Freud (1938) describes the period of infancy and early childhood as characterized by "infantile sexuality," which reaches its height during this third, or phallic, stage. He held that the apparent later loss of memory of experiences of the latency period was not "a real forgetting of infantile impressions but rather amnesia similar to that observed in neurotics of their later experiences, the nature of which consists of the memories being kept away from consciousness (repression)". He further maintained that "the influx of this [infantile] sexuality does not stop, even in this latency period, but its energy is deflected either wholly or partially from sexual utilization and conducted to other aims ... a process which merits the name sublimation."

At any rate, according to Freudian theory, the resolution of conflict and the relief from stress through repression and amnesia mark the beginning of the latency period, the fourth period in emotional (psychosexual) development. This period covers approximately the ages 5 or 6 to 10 or 11 years. In certain respects development does seem to be latent during this period. Anatomically, the reproductive system shows no appreciable growth. Although the body as a whole grows steadily in size, there is little change in bodily contours and features. Relatively little sex differentiation takes place.

Functional development both motor and cognitive, however, makes marked progress. A large number of locomotor and—manual skills are acquired and the child's interests and his range of information expand greatly. As a person there is steady development. During the five or six years of so-called latency a rather remarkable behavioural metamorphosis frequently takes place. The child is likely to change from a rather agreeable, compliant, talkative, and generally delightful individual to a loud, intrusive, and rather impudent preadolescent.

With the onset of puberal changes, the child experiences a period of physiological instability. During this fifth, or puberal period, there is much preoccupation with the body and its changes. A step-up in interest in the opposite sex and many other changes in feelings and attitudes make this period an extremely important one developmentally. In terms of Freudian theory, pubescence sets off a reawakening, a resurgence, of sexuality that for a time has been latent.

These five periods may be taken in broad outline to represent emotional development. Theoretically, at least, they are directly related to specific stages in the maturation of the organism. The oral period is one of infant helplessness. The anal period begins as the child, through the maturation of body structures, begins to overcome his helplessness, to operate with some degree of independence, and thus to have shifted to himself some measure of responsibility for his independent acts. He must begin now to conform to the cleanliness requirements of society by gaining control of his hitherto involuntary eliminative processes. The demands of society force upon him a preoccupation with anal functions during the period when maturation is in the process of bringing the body and those neural structures involved to a stage of development in which voluntary control is possible. The maturational basis for the early phallic period, however, is not so clear. Perhaps general maturation of the child's organ systems, particularly of the brain cortex and the neuromuscular systems, makes possible a shift from conscious efforts to control the eliminative processes, and a preoccupation with them, to an increased interest in his parents and their relationships with each other and with him. Hence, the development of the Oedipus situation, with its tensions and conflicts.

The latency period begins with the gradual cessation of the stressful genital period. Controls have been achieved, the pressure is off, and considerable independence has been gained. The child's tensions and gratifications are no longer so largely determined by the functions of the alimentary canal and his intense relations with his parents. They come now largely from a physically active life in an interesting world of things and people outside his home and family situation. The developmental changes that characterize the puberal period are quite dramatic. They furnish an obvious basis for the many changes in interests and activities, which take place during this period. Implications of puberal changes for emotional development and adjustment are profound.

A Development Sequence Based on Infant Observation

The Freudian sequence of phases in psychosexual development, as we have noted, has mainly to do with change in sources of gratification and objects of affectional attachment. For many students of personality, and particularly for those engaged in clinical diagnosis and treatment, this conceptual sequence has been found useful and meaningful, and has been widely accepted.

For many of the more objectively oriented students of emotional development, however, there is a need for more specific behavioural indicators of the onset and development of the various emotional behaviour patterns and affective experiences of daily life.

Perhaps the nearest approach to a useful series of developmental stages for the first two years of life comes from an investigation by Bridges (1932). In this study, based on observations of the emotional behaviour from birth of 62 children, a careful record was made in each case of a response and the environmental conditions that preceded it. One outcome of the study was a generalized description of emotional development in terms of four kinds of change, as follows:

1. Intense emotional responses become less frequent.
2. Emotional responses are transferred from one situation to another, which did not originally evoke them.
3. The reaction to any given stimulus becomes more specific.
4. Emotional responses gradually become more differentiated.

In relation to the fourth aspect of change, that is, differentiation, Bridges described, with approximate timing, the sequence that takes place from birth to 2 years of age. From the beginning certain kinds of stimulation give rise to a generalized, overall excitement, which Bridges regarded as the original emotion. Infants who are thus excited are described in terms of suddenly tensed muscles, quickened breath, jerky, kicking leg movements, and wide-open eyes gazing into the distance. This pattern was found to persist throughout the first 2 years of life.

At about 3 weeks of age the first step in differentiation takes place. Distress, or unpleasurable excitement, as distinct from generalized excitement, appears. In distress there is greater muscle tension, more interference with breathing, louder and more irregular crying at a higher pitch. Tears may flow,· the face becomes flushed, the mouth distorted, and the fists clenched.

Not until the age of about 3 months does the distinctly pleasurable sort of excitement, delight, become differentiated. This pattern involves kicking, faster breathing, opening of the mouth, crooning sounds, and smiling. In contrast with the distress pattern, the movements in delight are free rather than restrained, the eyes are opened rather than closed there are smiles rather than frowns, and approach movements rather than withdrawal.

The next differentiations in the sequence are on the unpleasurable side. According to Bridges, at about 5 months of age the baby shows unmistakable signs of anger, as differentiated from the more generalized distress. The clearest sign of anger is a wail, as if in protest, without the closing of the eyes. Closely following anger comes disgust, characterized by coughing, sputtering, and frowning while the baby is being fed. The third differentiation from distress, according to this study, is fear, observable at about 7 months of age. The presence of a stranger is the most frequent stimulus to fear. The behaviour pattern includes general inhibition of movement, tears, and steady crying, with eyes tightly closed and head bent. The body remains rigid and inactive.

At about the same time (7 months) a pleasurable emotional pattern, elation, becomes distinguishable from generalized delight. The baby smiles, takes a deep breath, and appears to express satisfaction in a sort of grunt.

By the eleventh month there are clear signs of affection for adults. The baby at this time will put his arms around the adult's neck and stroke or pat the face with obvious delight. Sometimes he will approximate a kiss by bringing his lips close to the adult's face. By age 15 months he makes an affectionate response to other children as well as to adults.

Very shortly after affection manifests itself, jealousy appears as a further differentiation of distress. It is characterized by tears, bent head, and by standing stiff and motionless. In some instances anger and aggression are part of the picture. Finally, at about 20 to 21 months, the pattern described as joy makes its appearance.

This sequence of differentiations described by Bridges perhaps comes nearest to meeting the requirements of a series of progress indicators in emotional development. However, as we have noted, it covers only the first two years of life. The different behaviour patterns are extremely difficult to recognize and distinguish in a baby's behaviour. A given child's status in emotional development, we must therefore conclude, must be estimated mainly in terms of a general understanding of emotionality and of the changes that take place with age, rather than in terms of a definite scale of progress indicators.

Maturation and Learning in Emotional Development

As was suggested in our discussion of the Freudian sequence, maturation as well as learning is involved in the development of emotions. Freud's psychosexual stages relate directly to bodily structures and their maturation.

Emotionality in general, however, since it is so pervasive, involving as it does the functioning, more or less, of the total organism, is not so closely bound in its changes with time to the maturation of a specific bodily structure as is the case, for example, with the development of mentality. The great diversity of emotional patterns and modes of emotional adjustment to be found among otherwise mature people must be accounted for very largely in terms of learning in environmental diversity. This, of course, is not to discount the fact that individual differences in constitutional nature have much to do with determining idiosyncrasies in reactions to given environmental situations. It appears to be true, then, that beyond the early period of affective differentiation, as described by Bridges, the role of the maturational factor in the direction and extent of emotional change is of relatively minor importance.

Planned and Incidental Learning

That significant changes in the quality and patterning of emotions do take place, particularly during childhood, there is no doubt. Earlier in this chapter the two opposing categories of emotional experience were mentioned, and in that connection we emphasized the importance of the prevailing quality-positive or negative—of the child's home environment in determining the general nature of his emotional and social attitudes and behaviour. More specifically, evidence suggests that the prevailing quality of the experience the child has with his mother during early childhood is of paramount importance.

I. D. Harris (1959), in an intensive study of a group of children and their mothers, concluded that a mother's loving care—her warm dependability and her understanding of her child's individual needs for expression—was "crucially" related to later desirable emotional and social adjustment and that the lack of such warmness and understanding was a factor in later maladjustment. In the words of Harris:

> The mother can be a source of nutriment and pleasure for an infant as he feeds, but she can also be a source of pain, tension and frustration. If she is a source of pleasure, then the growing child will look upon others, and later humans as gratifiers; if she is a source of pain and tension, the growing child will react to others and to later humans as frustrating.

During the toddler period, when the "socialization" of the child is underway in earnest, progress in learning takes place at a rapid rate. Two general types of learning influences can be identified during this period particularly. First, the child, of course, learns those patterns of behaviour—personal and social habits and ways of speaking and thinking—which the parents set out to teach him in order to make of him an acceptable member of society. But while this deliberate inculcation of the culture is taking place the child also learns things that were not deliberately planned or consciously taught by the parents, There is likely to be much incidental emotional learning taking place, For example, the child is likely to learn, by association, a joyful response or an indifferent one or one of dread and resentment at the appearance of his father. He may learn to react to his parents generally with positive, approaching affection, or he may learn to associate them with frustration, anger, or fear.

The sort of emotional learning that does take place, in any case, depends upon the quality of the emotional interaction that prevails between the child and his parents. And the quality of that prevailing interaction is dependent upon two sets of factors: *(1)* the particular personal make-up of each parent (predominant traits, attitudes, expectations, frustrations); and *(2)* the child's unique pattern of temperamental tendencies such as those discussed (another example of organism-environment interaction).

Imprinting versus Healthy Emotional Learning

There is another line of evidence from recent research, mainly animal research regarding the imprinting phenomenon, however, which suggests a somewhat different interpretation of the relation between learned emotional patterns and environmental stimulation. The suggestion coming out of these studies is that perhaps it is not so much the pleasant quality of the emotional interaction between parent and young child that makes for positive emotional attachment as it is that there be strong emotional content of some sort in the of some sort in the interaction during the early critical period. In Scott's words:

> It should not be surprising that many kinds of emotional reactions contribute to a social relationship. The surprising thing is that emotions which we normally consider aversive should produce the same effect as those which appear to be rewarding.... [This fact] provides an explanation for certain well-known clinical observations such as the development by neglected children of strong affection for cruel and abusive parents.

In view of apparent conflict in the evidence cited by such investigators as Scott (1962), J. D. Harris (1959), and Diamond (1957), it is clear that a distinction must be drawn between an early emotional bond due to the "imprinting process" and healthy emotional adjustment and development in children. E. H. Hess's (1960) ducklings did become the more strongly imprinted to follow the mechanical model duck the greater the effort arid the pain they encountered in following it. Harlow's (1958) infant monkeys did crawl desperately toward their rejecting, punishing mothers, who, in their own infancy, pad been reared on terrycloth dummy "mothers" instead of interactive monkey mothers. Childreri have been known to become emotionally attached to cruel, unloving parents. But Harlow's dummy-reared monkey mothers had not developed the normal monkey interactive behaviour patterns in their association with and close attachment to their dummy mothers. Furthermore, in the vast majority of instances, human individuals who have been reared by neglecting and cruel parents must overcome great handicaps in acquiring the capacity to interact warmly and affectionately with others. Scott (1962) summarized the evidence as follows:

> There are demonstrable positive mechanisms, varying from species to species, which bring young animals close to other members of their kind: the clinging 'response of young rhesus monkeys; the following response of chicks, ducklings, and lambs, and other herd animals; social investigation, tail wagging, and playful fighting of puppies; and the visual investigation and smiling of the human infant. These are, of course accompanied by interacting responses from adults and immature members of the species: holding and clasping by primate mothers, brooding of mother hens and other birds, calling by mother sheep, investigation and play on the part of other young puppies, and the various supporting and nurturing activities of human mothers.

These interacting responses, on the part of others, normally involve Pleasurable and satisfying emotional content and make the difference between simple imprinting of an emotional bond and the establishment of the foundation for healthy interpersonal relations and emotional adjustment. The capacity to interact warmly and creatively with other human beings is generally the product of early emotional learning in interaction with a warmly dependable and understandingly nurturant "mothering one." This kind of emotional learning is qualitatively different from, and more than, the simple imprinting of an emotional bond.

Love : A Process of Interaction

Throughout our discussion of emotionality, the active aspect of emotion-that is, the idea that an emotion is a function of the organism and that when one is emotional a complex process is underway has repeatedly come to the fore. We were also reminded of the two-directional nature of emotionality. There are emotions in which the direction of activity is away from or in opposition to the object of the emotion. These are the negative emotions. The action is individualistic, separating, alienating. One becomes angry at, fearful of, and withdraws from another.

Approach behaviour characterizes the positive type of emotion. One draws near to and acts in relation with another. Love is the prototype of positive emotion. Love is uniting, integrating, need satisfying. The action is interaction. Each of the two persons involved is both actor and object of the action, giver and receiver. The relation is reciprocal. The mother, for example, holds her baby close, thus gratifying their mutual need for closeness. Her acts of love stimulate in her child reciprocal love responses. The interaction is uniting and mutually gratifying. It tends to be what Erich Fromm (1956) called "interpersonal fusion."

Emotional Deprivation

The term "emotional deprivation" refers to a lack, in the child's experience, of a positive reciprocal relationship with another person. It refers either to the rather precipitous loss of an accustomed relationship with a nurturing person in which his need for warm, loving care has been regularly gratified or to the absence from the beginning of conditions that permit the formation of such a relationship. Available evidence suggests that both of these types of deprivation can have profound effects upon the development of the capacity to function in the reciprocal love relationship.

We have already noted the importance of adequate and appropriate stimulation for optimal development during infancy. Stimulus deprivation is perhaps the more appropriate term to apply to the situation of an infant lying by itself in an "institution" crib during the first few months of its life. At first, other persons, as persons, are of no significance to a baby regardless of where he may be. He spends much of his time sleeping. But when he is awake he is very responsive to stimuli both external and internal. He cries immediately at any bodily discomfort and is just as immediately comforted by being cuddled in warm and close contact with another person. He may respond with excitement to the sights and sounds of persons moving about him and particularly to being "talked to" and to the other noises people make to babies.

In the usual family situation, the baby does not want for stimulation. The baby is likely to get much patting, stroking, and fondling, particularly if there are other children in the home, and this attention along with the frequent feeding and bodily care he receives often leaves little waking time for him to be "bored." This kind of stimulation, of course, is what was lacking in the traditional "institution" as described by Goldfarb (1945), Spitz (1945, 1946), and others. Furthermore, the indications are that it is the lack of general stimulation, rather than the lack of a specific "mothering one," that is responsible for the retarding effects of institution care during the early months of the baby's life. It is likely also that this retarding effect is the result of the lack of opportunity to learn rather than an actual maturational retardation. In this connection Yarrow (1964), after a careful survey of the relevant literature, summarized the situation as follows:

The institution is not simply an environment lacking in a mother-figure with whom the child has developed an attachment; institutional environments tend to be deviant in many other respects, such as in the amount, the quality, and the variety of sensory and social stimulation, and in the kinds of learning conditions provided (Goldfarb, 1955; Rheingold, 1960, 1961; Provence and Lipton, 1962; David and Appell, 1962). The low caretaker-infant ratio is associated with significant deprivation in the sheer amount of maternal care. This quantitative deprivation in maternal care, in turn is associated with inadequate kinesthetic, tactile, social and affective stimulation.

Although studies from a variety of sources indicate that environmental deprivation may have serious effects even on very young infants, there are very few data on the important questions of how early in infancy separation from a mother figure begins to have an impact. Schaffer's (1958) research is one of the few studies with data directly relevant to this issue. On the basis of findings that overt protest reactions to maternal separation, such as excessive crying, fear of strangers, clinging to the mother, are not evident before 7 months of age, Schaffer concluded that separation reactions appear "relatively suddenly and at full force around 7 months of age". By definition, true separation reactions cannot appear until after a focused relationship with the mother has developed.

The baby soon begins to show signs of an emerging ability to perceive—to react differentially to objects as objects and to people as people. He begins to respond differently to different people. His affiliative interest increases, and he begins to relate to persons as individuals and to interact emotionally with them. The pleasures of gratification and of relief from pain and discomfort soon become associated in his experience with a particular person, usually his mother. A strong affectional bond thus becomes established between mother and baby. Thereafter, no amount of casual or impersonal stimulation can take the place of that person-to-person emotional relationship. From that point on, to have that relationship disrupted, and for the child to be denied dependable, intimate contact with the "mothering one," would be more than stimulus deprivation. The term "emotional deprivation" would, under those circumstances, more accurately describe the child's situation.

Research Concerning Emotional Development

Most of the early studies of maternal deprivation in institutions involved children who were separated from their mothers and placed in the institutions at ages over 1 year (Robertson and Bowlby, 1952; Roudinesco, David, and Nicolas, 1952; Spitz and Wolf, 1946). The reports of these investigations were practically unanimous in regard to the emotional damage these children suffered. The general pattern of immediate reactions to separation were crying and strong protest, followed by progressive withdrawal from relations with people. After the immediate protest and crying came signs of despair and resignation and apathetic behaviour. As a rule, these children formed no emotional attachments to any of the institution personnel and even showed very little feeling toward the parents when they visited them later on. They acted as if "neither mothering nor any contact with humans has much significance for them".

However, there is relatively little direct research evidence regarding the long-term effects of maternal separation. Certain follow-up studies have produced suggestive evidence. Bowlby (1944), for example, in a clinical study of a group of young thieves, characterized certain of 'them as "affectionless characters." An analysis of the backgrounds of these individuals indicated that they had been separated from their mothers in infancy. In general, however, results of these follow-up studies have been quite varied. Different children have different experiences in relation to separation, and, due to the diversity of temperamental nature among the children, these experiences have different meanings for different children.

●●

4

Role of Mental Health

Development Mental Health and maintaining optimal human functioning is a task never completed. Mental health, as we have seen, represents a style or process of living and social interaction that spans a lifetime. It is not a final stage or condition of a person to be sought or achieved by special training or treatment. Rather, mental health is a dynamic function or property of a specialized energy system-the person-that participates in many larger systems. It is important, therefore, to study how an individual develops within and interacts with these larger systems or social groups in order to understand how optimal human functioning can be fostered.

The child develops within a social matrix. The nature of that matrix influences what he learns and how he feels about it, even though the processes by which learning takes place may be the same in all societies. Each culture and, to a lesser extent, each group to which the individual belongs provides patternings of expectations and relationships which influence the development of the child's behaviours, skills, and attitudes. More generally, it is only by becoming a participant in society that the child becomes fully human, acquiring a sense of self and realizing his potentialities for symbolic thought.

In this chapter we shall review the growth of the individual as a system as he operates in and contributes to other systems. We shall examine how positive and negative functions of both group and individual systems originate, increase or decrease, and how these systems mutually influence each other. Crises typical of particular periods of development and how the individual meets them with adjective or mal-adjective behaviour will be reviewed. Such an analysis might reveal some of the factors responsible for mental illness and also might suggest methods for preventing some of these factors from having effects, and thus help us to promote optimal human functioning.

Significant Interactive Systems

At one and the same time a person may be a member of many different systems, some of which overlap or include other systems. Let us consider a particular young man named Andy. As a family member, Andy is part of a group system. Andy belongs to a church, but his father and sister attend a different church. Some of his brothers attend the same school as Andy, however. So we see that Andy does share group systems with some members of his family group, but not all of his family participates in the same systems. All of these groups are located in the same town, however, so one system does encompass all of them. Each system varies in its size, members, complexity, rules, and structure.

As a member of multiple groups, Andy plays a role, perhaps a different role, in each system. Each group, therefore, has a special meaning for him. When any group is important to a person, for whatever reason, it is called a significant interactive system. All through Andy's life changes will occur in his significant systems. As older systems add or lose members, change leaders or structure of relationships" new or reconstituted systems will become more relevant to Andy. As a child the family group was completely dominant in his life, but as he grew older he participated in other systems—his playmate, school, scout, church, and other groups. Gradually he became less dependent upon his family and related more to these other groups.

Meaningful Interpersonal Relationships

Within each system certain persons often seem to stand out and to play uniquely influential parts in the life of an individual. Such interactions are called meaningful interpersonal relationships (abbreviated as MIRs). These relationships, or MIRs, become vital to the development of the self. As we have seen the self is dynamic, not static or unchanging. It matures and changes just as any other bodily system-physical, intellectual, or emotional. Just as all development depends upon continuous and varying interaction with environment, the self develops in continuous meaningful interaction with people, their models and surrogates, or their roles as played in fantasy, ideals, or wish.

These MIR interactions commence very early in infancy and proceed to form the basis for affectional bonds and role identifications, first with parents, then with siblings, peers, teachers, coaches, employers, girl friends or boy friends, work associates, one's children, and many others. Such MIRs need not be positive, however. They can involve negative reactions and hostilities, as well. Often the persons whom we dislike and reject can contribute significantly to the self. When such attitudes predominate, the child may appear to be belligerent or aggressive.

Throughout one's life the patterns of one's MIRs change and reconstitute themselves, quite like the colors in a kaleidoscope, as one finds new persons, roles, or models that assume importance to him and those of former significance die, move away, or lose relevance. Along with changes in these patterns come changes in the self. These MIRs help to protect, strengthen, and stabilize the self by providing a dependable structure or supporting persons or models. In turn, the strength and security thus afforded allows a person to feel adequate and safe enough to reach out to acquire new MIRs, thus permitting further changes in self.

Just as the self system depends upon MIRs, so do significant group systems. The patterns of affection and influence within the family group affect not only the functioning of the system as a whole but the functioning of individual members as well. Great shifts in family authority patterns and interdependencies occur when a parent dies or becomes disabled. Divorce or family quarrels may radically alter the MIRs between family members, thereby disrupting a family system. There is thus a close association between group systems and their component MIR affinities or negative attitudes.

Stages in Development

Although we recognize that there are important differences in how and when changes in group and individual systems take place, it is more convenient to analyze these changes in terms of similarities among people. Each pattern of behaviour seems to have a life cycle of its own in which it begins, grows, flourishes, and then blends into a superseding pattern. We find such patterns in all subsystems of the individual: physical, mental, and social.

When a certain constellation of traits seems to characterize a given age group it is called a developmental stage. Piaget and Inhelder (1958) described certain stages of intellectual development, for example. From birth to two years is called the "sensorimotor stage"; from ages two to seven years the child passes through the "preconceptual" and "intuitive thought" stages. From age seven to 11, he is in the stage of "concrete operations," after which, until age 15, he can master "formal operations." Other stages in behavioural development have been noted for such diverse traits as learning moral rules of right and wrong (Piaget, 1932), psychosexual behaviour (Freud, 1905), concept formation (Werner, 1948), and self-awareness (Nixon, 1962).

Such complex changes in behaviour are orderly and continuous, one stage of development forming the basis for the next, and merging almost imperceptibly into a new pattern. Jenkins, Shacter, and Bauer (1953) have observed the following:

> All normal children will follow an essentially similar sequence of growth, yet because of the great variations in endowment and experience and the interplay between them, no two children, even in the same family, will pass through this sequence in just the same way.... within the range of "normal" some children will develop much more rapidly than the average, some much more slowly.... But in every group of children some will be ahead of the others of their age physically, mentally, and emotionally, and some will be behind the others in one or all aspects of development. So when we talk about "the six-year-old" or "the eight-year-old," we are talking of averages—the stage of development most children reach at six or eight. Some children will reach "six-year-oldness" at five or even four, others not until seven or eight or even later.

From this statement we see that all persons do not enter a stage at the same age, nor are all traits typical of a stage necessarily revealed in a particular person. The boundaries of the stages are quite fluid. There is great variation between persons.

Until maturity is reached, most developmental stages are closely associated with age and biological growth. This is true especially of physical and mental ability. After physical maturity has been attained, however, the stages are related more closely to events and environmental conditions. The behaviour pattern of the young adult, for example, depends much more upon his marital status, his employment, and his social group membership than upon his chronological age.

Developmental Tasks. At each developmental stage, certain skills, behaviours, and roles need to be mastered in order for the individual to adjust satisfactorily to his several systems. Such needed behavioural patterns are called developmental tasks. Some developmental tasks relate to physical maturation, such as learning to walk. Other tasks arise from cultural expectations of society, as, for example, learning to read. Still others arise from the personal values and goals of a person that are part of the self, such as forming a philosophy of life. Mastery of developmental tasks at any stage is vital to adjustment not only at that particular developmental stage but to future stages as well.

In America, for example, one developmental task of early childhood is to learn control of the elimination of bodily wastes. Should a child, due either to immaturity or to emotional stress, not be toilet-trained by the age of two, family pressures to conform will increase. The child who makes such "mistakes" may be punished or rejected by his parents or teased by his brothers and sisters. These treatments could have lasting effects upon his self-image and his MIRs.

Another example might be found in the middle childhood period. Suppose 'a boy has failed to learn the physical skills needed to play rough-and-tumble games, perhaps due to weak physical development, illness, over-protection, or fear. This might hinder his assuming a masculine role identity. In young adulthood he might shrink from competition or from taking a new job because he lacks self-confidence. It is important in promoting mental health and preventing later maladjustment for a person to master developmental tasks at the time they are required by society and when he is physically and mentally mature enough to perform them.

Critical Formative Periods. For all types of learning, there is a period of readiness when the organism is "ripe"—a time, which is most conducive to learning. Learning prior to or after this time is likely to be inferior, perhaps even impossible, depending upon the time intervening prior to or following the formative period and many other factors. Studies of certain kinds of learning, particularly those related to intelligence, language development, and social responsiveness have shown that it is difficult for a person to make up for or to offset deficient or inappropriate learning which has occurred during a formative period.

One study by Goldfarb (1943) compared the later development of children who had been reared in an institution for the first three years of their lives with children who had been raised in foster homes. In adolescence, the children who had been reared in the impersonal environment of the institution showed more problem behaviour, lower mental ability, and less social maturity. Even those raised initially in the institution who had later entered foster homes that were more intellectually stimulating and emotionally supportive continued to be retarded in mental development and to be socially callous. Apparently the lack of early stimulation of language skills and the impersonal, emotionally cold social environment during the formative period for learning language and social responsiveness contributed to a lasting impairment of the development of these characteristics.

Many aspects of system growth—mental, social, and emotional—thus have critical periods, which, if passed without the learning of certain tasks, handicap the mastery of developmental tasks that depend upon it. Should a male child, for example, fail to develop a secure MIR with his mother in infancy and early childhood, he may fail to transfer positive MIR attitudes to his father, such as a sense of trust. Later, because he does not accept his father as a model, the same young boy could experience role identification conflicts during the middle childhood period, with concomitant difficulties related to this in adolescence. In another case, a girl who learns how to get along with playmates—the usual give and take-in the middle childhood period when social skills are forming is more likely to find social relationships in her adolescence and adulthood satisfying and meaningful.

Significant interactive systems, such as the family, also show developmental stages and critical periods of growth. Just as MIRs with family members affect the individual systems of each participant, so do the patterns of MIRs affect the family system as a whole. Families that have close affectional bonds, or positive MIRs, meet crises, such as bereavement, as problems for the group. In families where MIRs are negative, involving hostility or jealousy, for example, such crises precipitate quarrels and separate, rather than group, task-oriented behaviour. There are critical periods in family system development which if weathered successfully and the family learns to cooperate as a team respecting the individuality of each member, the family survives and functions well as a system. But if parents cannot accept the increasing demands for independence of their adolescent children, or if young parents cannot agree on how to discipline the children, these may become critical periods, not only for the sorrowing children, but for the family itself. If the critical period for developing family solidarity passes, basic damage is done, no matter how matters are patched up later. The family as a system, the basic linking MIRs, and the participants themselves all become permanently changed.

This can be seen with even larger systems, such as a church. Should a church fail to bring young people actively into the fold at a critical period in their quest for faith, it may never influence them. Then, lacking MIRs with younger people, the older members influence the system to suit their needs, and the whole system gradually undergoes a change, which makes it less and less attractive to younger people. This has far-reaching effects, because adolescents who seek a philosophy of life as a developmental task thus become less likely to seek or to find meaning or identification in a religious system.

Aims of Prevention

Along with advancing years come new stages of biological, psychological, and social development, each of which ushers in new tasks to master, new problems to solve, and new demands from new systems. Failure to learn developmental tasks, whether in periods of rapid growth or in the adult years, tends to make a person more vulnerable to traumas, frustrations, and stresses at later stages. What might be an ordinary problem for a person who has been successful in earlier phases could become a serious crisis for one who has not developed competence or resilience.

Primary prevention of mental illness, or its obverse, promoting optimum human functioning, has the following aims:

1. Equipping a person to cope with the stresses typical of each developmental stage. The objective is to facilitate the mastery of developmental tasks by providing for training experiences which will permit the person to build a diverse repertoire of responses upon which he can draw when frustrated. This strengthens his resources for dealing with problems of life and averts the establishment of inadequate, inefficient, or inappropriate habits.

2. Assisting a person in coping with problems, not only to teach him, but to prevent destructive side reactions to stress and to provide psychological support until the person can cope with the problems successfully himself. Although it may in some cases be a matter of solving the problems completely for a person it can more often be characterized as "lending a helping hand." The comforting counsel of a clergyman in times of bereavement, the vocational information by a school guidance counselor, the work of a Big Brother

with a delinquent boy, the practical nurse who cares for a large family of children when the mother is ill-all represent intervention to relieve debilitating reactions to stress and to assist in coping with problems.

3. Removing, reducing, or obviating unnecessary or avoidable stresses or obstacles that weaken or destroy a person's ability to cope adequately with his environment at critical formative periods and at crucial stages in later development. An example in the middle childhood period might be the provision for straightening protruding teeth, which could be a source of self-devaluation or social handicap. Making emergency loans to persons who are between jobs, shifting a child to a less punitive teacher, or observing special prenatal diet precautions might be ways of avoiding stresses or obstacles to effective coping.

As suggested above, many persons, not only professional workers, may be involved in promoting mental health and adjustment. In Stage I, the prenatal and early childhood phase, which includes many critical formative periods, the emphasis is upon providing the best possible maternal care and child training. In Stage II, preventive functions are performed by schools, community parent-child guidance services, various professional and informal consultants, and advisers on medical, legal, religious, governmental, and economic matters. Prevention may involve person-to-person intervention in dealing with short-term crises, as with vocational counseling, or it may entail general programing for public education and community services to combat chronic forces like delinquency or mental retardation. It may aim at large systems, as in slum clearance or veteran's benefits, or at smaller systems, as with family financial counseling.

Whatever the method, the distinguishing feature of primary prevention is in its goal of increasing the positive functioning of the individual and reducing his negative functioning, thus decreasing the risk of his becoming mentally ill.

Problems of Prevention

Which of many factors is most significant in contributing to behavioural maladjustment? Which factors are not causes, yet are associated with disorders? Of the factors we know to be important, how can they best be controlled? Is prevention feasible or even possible? These are problems that must be considered before an effective program to promote optimum adjustment can be inaugurated.

Rarely do causal factors occur alone. Interaction and compounding of causes are more frequently the case. This, of course, multiplies problems of prediction of behavioural outcomes and of instituting preventive measures. Consider the matter of maternal deprivation, for example. We know from accumulated evidence that lack of motherly care and attention at a formative period with institutionalized children tends to be associated with adolescent adjustment problems. The evidence is not consistent, however, because not all institutionalized children show these reactions. It is possible that some children who lacked mothering made up for it with other relationships. Possibly those who showed bad effects had been left in the institutions instead of being chosen for placement in foster homes because of poor health, physical unattractiveness, or maladjustment, any of which reasons could contribute to later maladjustment. Problems such as these prevent one from making certain conclusions regarding the effects of maternal deprivation. Even so, enough is known about the dangers of maternal deprivation to warrant having serious concern for its possible dangers and to justify steps to avoid it, despite the fact that all persons might not be equally affected by it. Most behaviour problems have multiple causes.

Mere association of a factor with a disorder or a negative function does not mean, necessarily, that this factor has caused the disorder. Carefully controlled investigations are needed to establish the Causes. Even when such factors are known, however, other factors could modify or dilute their effects and thus obscure their possible influence. Because of the large chance of error in predicting individual behaviour, some mental health workers suggest that prevention is not only difficult but impossible or impracticable. Others perceive the role of prevention to be one of reducing the risk of mental illness for larger populations. They believe that better control of the known contributing factors, even If they are not exclusive causes, might substantially minimize the risk for a whole population. The tasks of prevention are geared to both ends, actually: *(1)* reducing overall incidence of disorders, and, hopefully, *(2)* forestalling individual maladjustments. As in all behavioural sciences, the uncertainties are numerous and great.

Tarjan (1961) summarizes the problems as follows:

Mental function is affected by a variety of forces. Some act singly, others in combinations concurrently or at different times; there are infinite possibilities for interaction. Repeated and relatively mild infections, for instance, may have a suppressive effect on adaptive ability and intelligence. The quality of mother-child relationship may influence the clinical manifestations of the sequence of birth trauma. Socioeconomic status may have an effect on the frequency of perinatal [birth] injuries.... Though present knowledge is limited the utilization of known facts in preventive programs would produce significant results.

With these cautions in mind, we shall explore in the next section those factors that are associated with maladjustments during Stages I and II, the predecessors of mental health or illness. Optimistically these factors might suggest how to minimize hazards to mental health and how to achieve optimum adjustment.

Guiding Mental Health in Stage I

At the instant of conception, a new life begins. From this moment until the child enters school around the age of six is perhaps the most important period in his life for building toward optimum adjustment in later years. His heredity, which determines in many important ways what he can become or fail to become, has "set" the course of his structural and functional development. Whether this inherited potential will be realized or thwarted, facilitated or distorted, maximized or curtailed, will depend upon the environments encountered by the growing organism. As one writer put it, "Heredity deals the cards, but environment plays them. The courses "set" by heredity are altered. constantly throughout one's life by positive or negative factors of nutrition, disease, accident, stimuli (or their lack), opportunities, and so on.

Stage I brackets these foundation years during which the fundamental physical structures and coordinating functions are formed. It is a span· of time with many critical periods—a time of dawning of basic physical skills, verbal proficiency, social relationships, and attitudes, and a sense of self. These skills and attitudes provide the groundwork for the developmental tasks of later adjustment, thus preparing the child for independent action in a world beyond his home.

Prenatal Influences. The first environmental encounters for the new being are in his mother's uterus. Suspended in a fluid world, the concept us receives his nourishment and chemical signals via the umbilical cord, by which he is attached to the placenta, a dislike structure attached to the wall of the uterus. He also swallows amniotic fluid in which he is suspended, thus receiving chemical substances present in the fluid. In this relatively stable environment the concept us develops very rapidly in a regular or orderly sequence from a small egg state, called the ovum, to a responsive fetus with many functioning organs in just two months.

Whatever interferes with his nutrition, oxygen supply, chemical balance, metabolism, or organ functions in the first three months, when the vital nervous system and brain are forming, may result in permanent malformation and subsequent disability, should the child survive. If the mother should contract rubella (German measles) in this first critical three months, for example, permanent damage to the fetus involving gross malformations or neurological deficit is highly probable.

Some representative outcomes of prenatal factors upon the subsequent mental health of offspring are listed in Table 4.1. The mechanisms of action in producing these results in many of the factors listed in Table 4.1, however, are not known. We are often unsure from the evidence whether a factor is directly responsible or whether some related variables could be at fault. For purposes of prevention it may not be necessary to make such distinctions, although we would have to admit that our approaches to prevention would be more direct and efficient, were the basic causes known.

The reader may note that mental retardation and neurological impairments predominate as outcomes listed in Table 4.1. Were fewer years to intervene between birth and onset of mental illness, which is more frequently detected in adult years, perhaps more definite associations of these prenatal factors and mental illness might be found. From present evidence there are too many confounding factors to be sure.

Malnutrition of the mother during pregnancy, passed on to the developing fetus as various deficiencies, is related to poor living conditions and inadequate food supply, often attendant upon war, depression, poverty, disaster, and ignorance. In one study of congenital malformations at birth in children of mothers from a very low socioeconomic population, Murphy (1947) found the rate to be 11.3 per 1,000 births, over twice the rate of 4.7 per 1,000 in the general United States population.

Table 4.1 Prenatal Influences Upon Subsequent Mental Health of Offspring

Factor	*Reported Outcome or Tendency*
Maternal Nutrition	
General malnutrition	Higher incidence of major and minor illnesses, neural defects, and premature birth (Burke, et al., 1943; Ebbs, 1942; Tompkins and Wiehl, 1954); greater risk of anancephaly (absence of brain) and deformities of central nervous system (Anderson, et al., 1958).
Iodine deficiency	Possible endemic cretinism (disorder with physical and mental deficiency).
Vitamin deficiency	Lower intelligence (Harrell, Woodyard, and Gates, 1955).
Maternal Attitudes and Emotions	
Rejection of pregnancy	"Neurotic infants": hyperactive, irritable, squirming, crying, unusually dependent, with feeding and sleeping problems (Sontag, 1944; Wallin and Riley, 1950).
Emotional stress and anxiety during pregnancy	Hyperactive, hyperirritable infants; mental retardation; "impaired motivation" (Sontag, 1944; Stott, 1959).

Contd.

Factor	*Reported Outcome or Tendency*
Maternal Dysfunctions and Diseases	
Diabetes (defective sugar metabolism)	Premature birth (see below).
Hypothyroidism (lack of thyroid)	Cretinism (see "Maternal Nutrition" above).
Toxemia (disturbances in blood circulation and kidney function)	Possible mental retardation (Battle, 1949)
Hydramnios (excessive fluid in fetal sac)	Anancephaly (absence of brain).
Rubella (German measles)	Deafness; microcephaly (small head, with retardation); malformations; high risk of mental retardation if infected during first 3 months of pregnancy.
Measles	Microcephaly
Influenza	Anancephaly.
Toxoplasmosis (infection of blood plasma)	Hydrocephaly (dropsy of brain with mental deficiency); microcephaly; mental deficiency; brain defects.
Fungus infections	Hydrocephaly; brain defects.
Rh factor (maternal-fetal blood incompatibility)	Brain damage; mental deficiency.
Maternal Age	
Under 20	Greater risk of mental retardation (Pasamanick and Lilienfield, 1955).
Over 35	Grea ter risk of anancephaly (absence of brain), defective spine, hydrocephaly (dropsy of brain with mental retardation), mongolism (a type of

Contd.

Factor	*Reported Outcome or Tendency*
	mental deficiency with distinctive physical deformities), malformations of central nervous system, mental deficiency, mental illness, and premature birth.
Order of Birth	
First born	More mongoloids in first birth rank at all maternal ages, except at 40 years and over (Smith and Record, 1955).
Fourth born and later	Greater risk of neurosis.
Other Conditions	
Anoxia (asphyxiation due to use of anesthetics, pain relievers, gas, difficult or prolonged delivery, etc.)	Mental retardation; neurological abnormalities; greater risk of cerebral palsy and. schizophrenia.
Smoking (excessive)	Premature birth (see below).
Premature birth	Higher incidence of mental retardation, preschool-age behaviour problems, "nervous disturbances," cerebral damage.
X-ray and other irradiation	Microcephaly; mental deficiency; hydrocephaly.
Birth injury	Various neurological and mental impairments.

It is encouraging to note that improved nutrition may improve mental status. Harrell, Woodyard, and Gates (1955) reported unusual improvements in intelligence of the offspring of women who had received vitamin supplements while pregnant. Children of mothers who had received varying amounts of vitamins prenatally showed an 8-point superiority in IQ over children whose mothers had received placebos containing no vitamins when tested at four years of age. The authors suggest, too, that had the supplements been given early, rather than late, in pregnancy the effects might have been even larger.

Numerous studies have shown that children born to mothers who were emotionally disturbed during pregnancy or who did not want their children produced hyperactive, irritable infants who had feeding and sleeping problems (Sontag, 1944; Wallin and Riley, 1950). Various glandular and blood infections and dysfunctions, such as rubella, are associated with such malformations as hydrocephaly (excessive fluid in the brain area), microcephaly (small head), and other brain defects, with mental retardation.

Many diseases and dysfunctions of the mother have been associated either directly or indirectly with brain defects or mental retardation in her offspring. The list includes diabetes, thyroid deficiency, blood disturbances (such as those due to Rh-factor incompatibility and infections like toxoplasmosis), influenza, measles, and fungus infections. Also, when the mother's age is under 20 or over 35 the risk of bearing defective offspring is greater. Other prenatal conditions found related to mental disorders or defects are excessive smoking (related to prematurity of birth), irradiation by X ray or other means, premature birth, injury in birth, or temporary anoxia caused by anesthesia, pain relievers, gas, difficult, or prolonged delivery.

Certainly an adequate prenatal preventive program should include measures that would seek to minimize the incidence or effects of these factors. Such programs might include improved prenatal medical care, education, and counseling of girls and brides concerning the importance of nutrition and the possible effects of certain prenatal factors upon their future children. Programs of economic and health assistance for the disadvantaged members of society would also be necessary. In order to demonstrate overall effectiveness and improvement resulting from preventive measures a large-scale approach that would reach the bulk of the population would be needed.

Family Influences upon Socialization and Adjustment. Five natural periods for important changes in socialization have been described by Scott (1963). Each constitutes a critical period in the development of adjustment. They are:

1. Neonatal period. (0-40 days of age).
2. Primary socialization (about 5 weeks to about 7 months of age).
3. First transition period (7 to about 15 months).
4. Second transition period (about 15 months to 27 months).
5. Period of verbal socialization and the struggle for autonomy (27 months to 4 years).

   We shall add one more period, which may be identified in this Stage I:

6. Period of development of identification and role (4 to 6 years).

1. Neonatal Period. During the first, neonatal, period, which extends from birth until an age of about 40 days, little or no socialization occurs. Physiological processes relating to sleep, feeding, and elimination predominate. The infant displays strong demands to use the few responses at his disposal, consisting of swallowing, sucking, crying, sneezing, yawning, and so on.

Although he does respond vigorously to pain and discomfort, the only distinguishable emotion is that of general excitement. With reasonably prompt satisfaction of demands and release from pain, there are no noticeable after effects. Although it is possible to establish conditioned responses of sucking, winking, and other motor reflexes within this period, it is difficult to do so, and the resulting conditioning is' quite unstable. Parents need not fear the emotional results of upsetting experiences at this time. The infant not only is incapable of experiencing intense emotions as adults might feel them, but he has a very short memory. Even social smiling responses are absent in this period.

More important, perhaps, than disturbing experiences of the child are the reactions that parents might have to such experiences. The crying of the child sometimes arouses feelings of empathy, guilt, or resentment in the parents, promotes quarreling, and creates tensions that disturb the nursing routines, and thus sets off a chain of psychological reactions that could persist to affect the marital relationship and the future attitudes of the parents toward the child.

Scott (1963) summarized the neonatal period as follows:

> ... the chief needs of the neonatal infant in the first 2 months are for normal physical care, feeding, and comfort. Since the neonate is not a completely self-regulating organism, he will need adequate handling and stimulation, or "tender loving care." Physiological stress is likely to be more dangerous than emotional distress, and the baby can be passed from hand to hand and carried from place to place with little disturbance if physical care is adequate.

2. Primary Socialization. The second period, primary socialization, begins at five to six weeks and extends to about seven months. Within this period, the infant can make associations with visual stimuli and smile in response to any human face, even to a Halloween mask. It is a time when anyone can approach and develop a social relationship with a child. Since the strongest relationships formed are with those he sees most often, reaching a peak at four to five months, the mother becomes the chief agent of socialization, although the father and other siblings may also assume important roles. Throughout this period fear responses to strangers increase, and the primary social relationship with the mother becomes stronger.

It was once believed that the increasing depth of the mother attachment was due to her association with the satisfaction of many of the child's basic motives, such as obtaining food or relief from discomfort. Recent research by Scott (1963) discloses that the attachment does not depend upon food intake, reward, or punishment. Rather, the primary social attachment seems to occur regardless of external conditions of reward and punishment, and is associated only with emotional arousal.

The infant also forms attachments of a similar kind to objects, such as stuffed toys, and places, probably as a result of the same process. Although an adequate physiological or psychological explanation for this process remains unknown, it is quite likely that this period of primary socialization represents the formative period for the basic MIR of the child with its mother.

Erickson (1963) suggested that the sense of trust emerges at this time. When there has been maternal deprivation at crucial periods in the six-to-nine-month period, lasting emotional reactions are likely, involving, among other responses, inability to relate to or to trust others and emotional blunting, as revealed in studies of neglected institutionalized children. Whether such results are due to lack of "mothering," sensory deprivation, or other factors, the matter has been debated, because the studies had weaknesses and uncontrolled factors. Some have argued that adverse reactions to separation from the mother by reason of institutionalization prior to seven months of age are due to sensory, or perceptual, deprivation, whereas psychological damage after this time, when the MIR has become stabilized, is best characterized as maternal deprivation (Schaffer, 1958). In either case, long-range studies are needed to determine the extent to which psychological problems considered related to early deprivation persist and the extent to which they could be reversed by environmental improvements.

Prominent among the experiences of the child during this period of primary socialization are the acts associated with feeding. Psychoanalytic theories of personality have stressed the importance of breast feeding, rather than bottle feeding, self-demand, rather than rigid feeding schedules, and gradual and late weaning as being essential for subsequent mental health. After an extensive review of research pertaining to infant feeding practices, Orlansky (1949) concluded, however, that no specific infant nursing discipline had a consistent impact upon the child and that "the effect of a particular discipline can be determined only from knowledge of the parental attitudes associated with it, the value which the culture places upon that discipline, the organic constitution of the infant, and the entire socio-cultural situation in which the individual is located."

Many factors are associated with the mother's choice of nursing procedures that might be more important than any specific nursing practice. Watson (1959) suggested that the general attitudes of the mother toward her child and the larger family pattern in which the nursing procedure is embedded are of greatest importance. A cold, rejecting mother may carry out such "approved" practices as breast feeding, self-demand feeding, and gradual weaning merely as a duty, whereas employment or illness may force an acceptant, affectionate mother to use bottle feeding on a schedule with early weaning. Regardless of specific practices, when feeding is accompanied by warm maternal attitudes and close physical contact between mother and infant, thus giving the child feelings of contact comfort and emotional support, the basic mother-child MIR is strengthened. The mother benefits as much as the child. To the extent that breast feeding, demand feeding, and gradual weaning foster this sensory stimulation and feeling of security without arousing conflicting maternal attitudes, they are desirable practices.

It is clear that in this period of primary socialization occurring between six weeks and about seven months the baby forms his most important social relationships with his parents and family members, particularly with his mother. Whatever disturbs the formation of these basic MIRs—separation due to hospitalization or institutionalization, lack of warm, satisfying contacts with a mother or mother substitute, or negative maternal attitudes—may have lasting effects upon the child, thus hindering his mastery of later developmental tasks which depend upon these MIRs and feelings of security and trust.

3. Frist Transition Period. Two distinct periods of transition are said to occur between the ages of about seven and 27 months, during which the child develops skills that facilitate his assumption of a more responsible family role. The period of primary socialization does not terminate abruptly at this time, however. Close family ties formed in the earlier period continue to deepen, and relationships with strangers become more and more difficult to form. There is considerable overlapping of periods.

During the first transition period, initiated by the eruption of teeth and crawling at about seven and a half months, the child becomes weaned from breast or bottle to a more adult style of eating. With greater mobility and increasing capacity to learn, he is able to respond to a wider range of stimuli, thus to become prepared for an adult mode of locomotion, walking, at about 15 months.

4. Second Transition Period. Walking marks the beginning of the second transition period, which extends to about 27 months. During this period the child begins to use and to understand words. "Do's" and "don'ts" become attached to many of his actions and form the basis for learning simple concepts of right and wrong, an important developmental task for this age. The active curious child is told to be quiet and to keep away from fascinating objects like electric outlets and china figurines. Even though he enjoys elimination processes, he is punished or scolded for performing when and where he should not. He must spit out interesting objects he finds on the sidewalk and is forced to give up or to avoid many activities that would gratify his motives and to acquire voluntary control over bowel and bladder processes, which are initially involuntary processes. It is a time of great frustration and feelings of conflict for both the child and the parent.

Excessive use of aversive stimuli by parents may spread negative reactions of fear and stress to many of the behaviours and persons in the world of the child. When a parent is angry or punishes a child for acts over which the child lacks control, as in the early stages of toilet training, emotion stemming from frustration in the child may actually interfere with his forming desired habits. Instead of learning to control his sphincters, the child learns to avoid and to fear painful stimuli and the loss of love and approval by his parents. He leams also to associate unpleasant stimuli and emotions with the whole process of elimination. Approach responses to master tasks and to please parents conflict with avoidance responses, thus further promoting stress and defense-oriented behaviour. Prominent among these defense reactions are those of hostility and aggression, wherein a child expresses defiance of parental wishes and anger through tantrums and rage. Such behaviour often arouses aggressive retaliation by parents by means of punishment and rejection. Anticipating such reactions, the child learns to fear and to inhibit his own anger and aggression, giving rise to what has been called anger-anxiety conflicts (Dollard and Miller, 1950).

Dollard and Miller (1950) described the origin and implications of these conflicts as follows:

> Anxiety responses ... become attached not only to the cues produced by the forbidden situation but also to the cues produced by the emotional responses which the child is making at the time. It is this latter connection, which created the inner mental or emotional conflict. After this learning has occurred, the first cues produced by angry emotions may set off anxiety responses, which "out-compete" the angry emotional responses themselves. The person can thus be made helpless to use his anger even in those situations where culture does permit it. He is viewed as abnormally meek or long-suffering. Robbing a person of his anger completely may be a dangerous thing since some capacity for anger seems to be needed in the affirmative personality.

One can see from the above that severe and strict methods of discipline, especially if initiated before a child is mature enough to comprehend or to master a task, may have lasting consequences. A whole cycle of punishment, fear, conflict, retaliation, punishment may commence which generalizes to the whole parent-child relationship in all areas of socialization, not only the specific cleanliness training.

The whole pattern may be accentuated if a parent who is punishing or rejecting has already acquired other anxiety-arousing characteristics.

If the mother has been nurturant during the first year of life, her positive value might be sufficient to neutralize some of the negative feelings produced by socialization demands without a marked change in the perception of the mother. But if the mother has been cold and rejecting, then use of strict toilet training is much more likely to have a deleterious effect on the child, and lead to negative feelings toward her.

Furthermore, a child who has received constant disapproval and rejection for his behaviour may come to regard himself as dirty, loathesome, worthless, unwanted, and unloved—hardly a feeling conducive to positive efforts to learn and to improve.

The above discussions suggest several points to consider for child training in the transition and later periods of developments:

1. Training in a socialization task should be begun only when a child is mature enough to master the task. Although training can be forced before this time, it is often at the expense of the child's feelings of emotional security and the formation of close MIRs.

2. No disciplinary measures should be employed that weaken or threaten to weaken MIRs. Parents should avoid making their love contingent upon the behaviour of the child. The child should never be made to feel he is unwanted, unworthy, or unloved, no matter what he has done. He needs to understand what he has done wrong and what he should do to be correct. He must learn also to accept aversive stimuli that will assist him to discriminate desirable from undesirable responses. But the aversive stimuli must not be so severe as to produce stress; it should be associated with the undesirable behaviour, not with the person producing it, nor with self-evaluations. Punishment should not interfere with the formation of healthy self-attitudes or block positive strivings to learn.

3. The aim of discipline should always be to promote learning. Prompt, sure, but light discipline insures learning better than delayed, inconsistent, or severe discipline. Discipline is not synonymous with punishment, however. Good discipline is corrective and strengthening, rather than merely punitive.

4. Appeal by parents to positive, rather than aversive, motives is more effective in the long run. This prevents generalization of avoidance responses to other areas of behaviour. A child scolded for not eating a certain food may

come to dislike mealtime, for example. A good rule is to accentuate the positive."

5. Many behaviours once regarded as "bad" by parents are now seen as natural and to be expected of children at this age. Such activities as thumb-sucking, bedwetting, or masturbation, in most cases, will cease to be problems without harmful aftereffects as the child matures, if undue attention is not paid to the behaviours. Severe punishment by parents tends to increase whatever emotional tension may have contributed to the problems, thus delaying the normal abandonment of the practices, sometimes indefinitely. It is better to try to understand the causes of the behaviour and to remedy them than to try to suppress or to scare the behaviours away by aversive acts which tend actually to fixate behaviours and to cause fear and hostile reactions that can generalize to the parents and to the whole home environment.

5. Period of Verbal Socialization and the Structure for Autonomy. As the child develops language a new type of social control and learning becomes available to the child. The stage of verbal socialization commences around the age of 27 months, according to Scott (1963), when the child becomes able to use and to understand sentences. Simple communication is available before this time, of course, but now the child is able to use and to understand more complicated rules and to assume greater responsibility. As he explores his world and experiments with his newfound powers, he expresses his individuality and learns to cope with many new situations, free of parental direction and help. Again, as noted with cleanliness training, the reactions of the parents can shape attitudes either of confidence or of over dependency of the child.

Such over-shielded and restricted children tend to become socially inhibited, fearful of trying new tasks, and excessively dependent upon mother and others for care and for making decisions. Although no parents want to expose their children to danger, if a child is never permitted freedom to try things on his own or to explore without fear of being punished or hurt, later developmental tasks that require courage, confidence, and spontaneity will be difficult, perhaps even impossible. The child will lack a sense of initiative to tackle new tasks and will remain immature and over-dependent. For example, it is important for the young boy in middle childhood to acquire skill in rough-and-tumble games.

Parents need to relinquish controls judiciously and to encourage the child to assume responsibility for his behavior commensurate with his ability to accept it. As soon as he is able, the child should be encouraged (not forced or coerced) to dress, feed, and wash himself, and to manage his own toilet functions. Boundaries of play areas can be relaxed gradually as the child shows he can accept responsibility. Children are usually eager to perform these tasks, to extend boundaries, and to acquire independence in the second and third year of age.

Even before he seems ready, the youngster may refuse help, resist rules, and generally insist upon having his own way. It is the beginning of a sense of self-esteem or pride in accomplishment. These periods of negativism, which will recur throughout the child's development, are expressions of his needs for mastery, his growing sense of pride in his skills or need for them—a struggle for "selfhood." Negativism becomes particularly pronounced at certain times when the child is experiencing difficulty in acquiring skills. Sometimes, the best help from loving parents as he learns these tasks is the least help. Much of the contrariness and pugnacity expressed by the child is due to the natural frustration he experiences before he succeeds in constructive problem solving. For a parent to provide too much help, too strong control, or to punish or to restrict the accompanying negativistic behaviour severely may compound the problems, prevent the development of frustration tolerance, and retard the learning of developmental tasks.

Parents need great patience, faith, and perspective to balance between crushing the child's drive for autonomy by forcing their will upon him and permitting the child completely free rein, which can become a "reign" over the household. A tolerant middle course is needed. The exasperated parent needs to reassure himself often: "This too shall pass!" Extremes of either punishment and rejection or permissiveness and indulgence are undesirable for both the family and the child. It is a critical period of socialization for both parents and children, which requires wise judgment, emotional control, and forbearance.

6. Development of Identification. In the period between ages four and six the child participates more and more vigorously in his environment. It is a period of enterprise and imagination when the child develops a strong sense of possession, becomes competitive, and extends his self to include "my house," "my tricycle"—everything becomes "mine." It is a most egocentric period because the child extends self by absorbing new territory (Allport, 1961). He also includes people in his orbit, so his attachments at this time can involve possessiveness and jealousy of his friends or siblings.

Should a new child be brought into the family about this time, the sibling rivalry can be disturbing to the whole family system. Parents need to be especially careful not to ignore the feelings of the older child when the new baby enters the household. Despite all assurances, the older child often perceives the threat of the newcomer to his basic MIRs. He finds MIRs very difficult to share. Such rivalry is strongest when the age separation is from two to four years, because it is then that the oldest child is developing his self-image and identifications.

One of the chief sources of a favorable self-image at this stage of development is the identification of the child with the role responses of significant others. Identification can be defined as a belief that the characteristics and attributes of a model belong to the self. Not only do parents and others act as direct rewarders or punishers of behaviours and as mirrors or interpreters of the child's behaviour, but they also become models for the child.

What determines the adoption of a model? One opinion is that adults command positions of power over their environment, possess competence and skill, autonomy, and have respect from others—all highly desirable goal states for the child. In order to share vicariously in these highly desirable goals possessed by the model, the child adopts beliefs and behaviours that he sees or believes are practiced by the model (Kagan, 1962). Another view (Mowrer, 1950) is that the child will value and imitate those behaviours that are characteristic of the persons he values most. The stronger the MIR, the closer the child will identify.

This is the period of male and female role identifications that play so important a part in later peer-group adjustments and the roles played as marital partners and as parents. The boy is expected to be masculine, to prefer the activities and interests of boys. The girl is expected to imitate the mother, to play with dolls, to play house, to enjoy dresses, to be interested in her appearance. When parental absence, illness, or overprotection prevents the formation of appropriate role identifications, the effects upon later role identifications may be extensive. Parental identification has been found important in the later development of vocational interests (Henderson, 1958; Stewart, 1959; Steimel, 1960; Crites, 1962), conscience (Kohlberg, 1963), and aggression (Mowrer, 1950).

Identification in this period contributes also to the acquisition of a conscience and moral standards. Whereas earlier, before his understanding and retention of language were well developed, he might respond to things he knows he has done wrong with fear of rejection or punishment by parents, now he can identify with how his parents might react and feel guilty. Because he is better able to anticipate consequences and to delay satisfactions, he comes to demand of himself the rules of conduct, which have previously been set and enforced by his parents. Identification is thus an important step in learning to conform to the rules and expectations of society. In his later years the child will interact with others and experience gradually the role identifications of others and come thus to appreciate the more complex moral concepts of reciprocity, justice, and group welfare (Kohlberg, 1963).

Patterns of Family Influence. Not too long ago people believed that strict discipline was the most important means of controlling the growing child. "Spare the rod, spoil the child" was a maxim that reflected this belief. Another maxim, "As the twig is bent, so inclines the tree," implied that early application of strictness was needed to insure the firm implantation of habits of obedience and "good" behaviour.

That strictness and strong discipline might have been successful in keeping children "seen, but not heard" was probably true, but we know today how many undesirable side effects are likely when discipline becomes the tool of a hostile and rejecting parent. We know, also, that children are influenced by many other factors in the home, such as the attitudes that accompany discipline, the patterns of alignment-that is, the connectedness and significance to the child of each other family member, including brothers and sisters, even grandparents. Often the total family situation provides a "theme" that sets boundaries to a child's behaviour and goals, which is even more effective than the interaction he has with any one person in the family circle.

Such patterns of alignment and "themes" often reveal the salience, or particular importance, which a family member might have—the MIR, which will determine how great an influence this person will have in shaping the child's behaviour. Each of us can recall how an older brother, Or perhaps an aunt, was important to us in approving or disapproving, punishing or praising, our actions.

Effects of Parental Attitudes and Controls. Most studies of discipline and control within the home have shown that the specific methods used are not as important as the attitudes that accompany the practices. Several dimensions of these attitudes have been studied. Affection or warmth, in contrast to rejection or hostility, has been given much attention. It has been found that when a warm and loving attitude prevails, the parents tend to control mainly by praise and reasoning, as positive actions, or by isolating the child, showing disappointment, or by withdrawing love, as negative actions (Becker, 1964). Providing love contingent upon behaviour is a highly questionable procedure. However, manipulation of affection tends to result in forming strongly internalized reactions in children, with guilty feelings when they sense they have done wrong and with acceptance of responsibility for their own actions. When combined with permissiveness, parental warmth tends to develop a child who is socially active, outgoing, creative, independent, and generally responsible. When combined with strictness of control, which is sometimes associated with overprotection, the child may be somewhat inhibited, submissive, dependent, polite, and not outgoing, friendly, or creative.

Power-assertive methods tend to be used by parents who are hostile toward or rejecting of their children (Becker, 1964). Such parents employ more physical punishment, threats, shouting, and coercion. In general, such discipline tends to induce hostility and aggression in children. This is often shown as disobedience and resentment of persons in authority, such as teachers, parents, or camp counselors, fighting and quarrelsomeness with other children, and tendencies to externalize their reactions to wrongdoing by blaming it on others, fearing punishment from others, and displaying hostility toward those who detect them, rather than feeling shame or guilt. Thus the aggressive, hostile parent often "reaps as he has sown" an aggressive, hostile offspring who is likely to experience continual difficulties in his social relations with both children and adults at later developmental stages. Such a pattern can be exaggerated when one parent, usually the father, is extremely hostile and rejecting, and the other parent is very permissive, or lax, about discipline. In such a situation, the hostility in the child, borne of his feelings of helplessness and frustration in the face of his father's brutal aggression, has no inner controls, and the result is often delinquent behaviour.

Boys and girls tend to respond differently to parental controls. One study has shown that boys who were given moderate but not harsh discipline from fathers and warm nurturance from their mothers showed the most responsibility and leadership. Girls, in the same kinds of situations, tended to overreact emotionally to much discipline, and to become more inhibited, anxious, and dependent. With girls, lighter discipline was needed to achieve the same results. On the other hand, lack of discipline from fathers who rejected or neglected their sons tended to produce irresponsibility and poor leadership in boys.

The optimal development of responsibility, leadership, and social outgoing characteristics with sound emotional development in children seems to be associated with parents who are typically warm and permissive, rather than rejecting, hostile, and restrictive. Such parents tend to use praise liberally and to reason with, rather than physically punish, their children. They use more positive measures, rather than using isolation or love withdrawal, or threat. Punitive or "contingent love" methods tend to inhibit behaviours rather than to reinforce and to strengthen the better social responses and acceptance of responsibility. This is not to say that discipline should never include punishment. Perhaps the optimal situation for child rearing and successful socialization is found in families *(1)* where boys receive light to moderate discipline from fathers and unqualified love from both parents, and where discipline is lighter for girls under the same conditions, *(2)* in which permissiveness, rather than restrictiveness, prevails, and *(3)* in which there is agreement rather than conflict between parents regarding discipline.

No rules can apply invariably, however. As Becker (1964) said:

> There are probably many routes to being a "good parent" which vary with the personality of both the parents and children and with the pressures in the environment with which one must learn to cope.

Guiding Mental Health in Stage II

As the child enters school around the age of six, he moves into Stage II. His basic, meaningful, interpersonal relationships with family members have been formed, and his role and status within this system have been defined. He has developed routines and habits expected by society concerned with walking, eating, elimination, and modesty. He has learned to communicate with others, to follow directions, and has formed simple concepts of right and wrong, the rudiments of a conscience, and moral character. His concepts of self now include a sense of identity, autonomy, self-esteem, and initiative. The child who has mastered these developmental tasks will not find this transition to Stage II difficult. He will be able to take in stride, without undue stress, his rapid physical changes, the expansion of his social world, and the demands for new skills and achievements in the school and on the playground.

Middle Childhood Period. In the middle childhood period, which ranges roughly from ages 6 to 12, the most demanding adjustments may be considered under three main headings: social, skills, and self. Speeded by new school relationships, the socialization process begun at home continues. At this point, parents often feel somewhat helpless they see their children, once their concern alone, being influenced and pending upon other people: teachers, playmates, group leaders, babysitters, and others. The best thing they can do is to try to maintain contact, to keep informed, and to supply the interpretations needed to foster healthy social growth. If there is continuing training in consideration for others, in taking constructive approaches to problems, and in good social and personal habits, the going will be smoother, but naturally we could not expect this period to be completely free of the kinds of stresses that are typical of any processes of change and growth.

At this stage, the principal role of the family is to strengthen the appropriate learnings and attitudes being formed by many experiences. Parents still playa significant role in defining the meaning of experiences for a child and the social roles he will play. Through their interpretations the child will develop basic understandings of competition, money, and social class distinctions based upon economic level, occupation, religion, and nationality. Attitudes and values regarding sympathy, achievement, respect for the opposite sex and for people in authority, alcohol, smoking, sex, honesty, justice, and loyalty originate largely in the family system.

It is during this middle childhood period that parents for the last time will play a dominant role in the education of their children. It is a stage in which parents are being taught the most by their children, too. Because of this; this period is a critical transition zone. With their power and strength to dominate, it is possible for parents not to recognize their roles in contributing to the insidious problems of rejection, over-control, or neglect, which may erupt later, full-blown as adolescent rebellion or severe self-devaluation.

Both at home and at school great stress is made upon achievement and skills. This can be a great source of anxiety and frustration 'for the child. Children differ greatly in their rates of mental, social, and physical maturation, and at anyone time a child may be at very different levels of development in each of these areas. This may cause a parent or teacher to regard a child as being a "problem," when all he may need is more time to grow. Although Mary might be tall for her age and faster in reading, she might be emotionally or socially immature. It is often difficult for a parent or teacher to know just how much to expect of a child. It might be too much or too little. The inconsistent demands and punishments of parents sometimes reflect this uncertainty.

Subconsciously, most parents want their children to achieve well in school, if possible to surpass their own degree of success. These desires tend to induce pressures upon the child to excel, which, if combined with rejection or punishment for lack of success, could affect seriously the child's feelings of self-esteem. In the peer group, where his social acceptance hinges more upon physical skills and strength, he may feel such inadequacy even more strongly. Caught in a cross fire of striving to win acceptance from both parents and peers, whose values may be opposed, the child might resolve the conflict by choosing to pursue the anti-intellectual goals of his age mates, much to the distress of his more school-oriented parents.

Parents could minimize such conflicts by reducing the emotional penalties associated with strongly negative controls, such as restriction, punishment, and rejection, which tend to intensify the aversive reactions of the child. Emphasis should be upon positive methods by building up his skills and confidence to meet problems constructively, in the knowledge that he will receive acceptance and approval for trying and for progress, not only for victory and full achievement. Pressures of strong competition tend to reinforce feelings of failure and inadequacy in a child who lacks the skills or maturity to become a winner. The child who feels confident that he will be accepted and will be loved whatever the outcome will be more likely to exert his maximum effort and to develop better frustration tolerance than one who senses that "all will be lost" if he fails. He also will be better oriented toward accepting the values and attitudes of his family. In his peer group he will also win acceptance because he is not a "quitter" and supports the group, rather than showing hostility and aggressiveness when he loses.

Friendships become increasingly important during the middle childhood period. The youngster has begun to make his own choices, and the group becomes a testing ground for his need for achievement, belongingness, and self-esteem. For boys and girls alike it becomes increasingly important to have a pal or best friend. Secret clubs, with special meeting places and secret codes, often make these relationships secure from others by excluding them. It is an important step in the development of those social bonds of loyalty and affection that precede the attractions to the opposite sex during adolescence.

Adolescence. The adolescent years, between the ages of 12 and 21, show very rapid social expansion. It is the era of the peer group. The adolescent spends most of his day, whether in or out of school, with others of his own age, so it is natural to expect this peer society to dominate his thinking and his behaviour. At a time when the adolescent is struggling for independence from parental controls, when he feels the greatest pressures of competition, inadequacy, and parental disapproval, the youth can find in his peer group freedom from a demanding adult world, opportunity for acceptance and even status, and secure MIRs. Thus the peer culture becomes both an arena and a forum for self-expression exploring new social roles, learning new standards of behaviour, and developing social skills in a group away from the family. Not all such learning is bad. At best the youth is afforded opportunities for status, responsibility, and leadership and for learning how to compete, to cooperate, and to share in group goals. Only when the peer group promotes antisocial. attitudes of prejudice, delinquency, or separates the youth from the beneficial influences of his home before he has absorbed them would there be serious social implications. Equally unfortunate is the severe social selection process of the peer group which can leave emotional scars upon the girl who is not accepted into a sorority or the school elections which are basically popularity contests.

Erikson (1963) mentioned important differentiations of self, such as the sense of identity, which emerge at adolescence. The young person begins to ask "Who am I?" and "Where am I going?" As he commences to separate himself from parental and home MIRs and to associate more and more with his age mates in new MIRs, he carries with him lingering doubts about his adequacy to stand alone. Once he was sure, but now he is not; he has found neither. a sure reason for being nor a tangible goal. He feels a need to be independent of his parents, but he still needs their MIRs. He feels many inner conflicts and anxiety over role confusion.

As the adolescent sexual functions develop, it follows that sexual tensions would contribute to such anxiety, especially if there were lingering uncertainties concerning masculinity or femininity of role. At this point, the adolescent seeks stability through new MIRs which will bridge the gap and stabilize the self. He may seek them symbolically in his religious faith, in new identifications with persons he can respect and admire (perhaps a hero astronaut, athlete, teacher, counselor, or coach), in peergroup pals, in "going steady," or even in early marriage. Where a close person does not gratify these needs the youth may display attachments to objects or symbols, as in fantasy, crushes on teachers or TV celebrities, love for pets, collecting class pictures and autographs, and so on. Many adolescent dating problems center about being "in love with love"—almost like being all dressed up and having nowhere to go. Yet it is from such MIRs as these in the adolescent years that firm loyalties to school, to teams, to clubs, which later develop into more remote allegiances to church and country, are formed.

Perhaps the most serious adjustment demanded of the adolescent is that of choosing and preparing for an occupation. With some 23,000 occupations to choose from, the adolescent needs to know a great deal, not only about jobs, but about himself, and to see these in relation to many variables, such as family expectations, financial limitations, opportunities for training, military obligations, automation changes of the future, going steady, idealistic versus realistic goals, and so on. Many of these conflict with each other, and give rise to many psychological defenses and emotional reactions.

The major tasks and adjustments to be made by adolescents are summarized in Table 4.2.

Table 4.2. Tasks and Adjustments of Adolescence

1. *Social maturation:* achieving emotional independence from parents and emancipation from home control; developing new relationships with age mates of both sexes; desiring and achieving socially responsible behaviour.

2. *Role development:* accepting one's physique; establishing a masculine or feminine role; forming a sense of identity.

3. *Occupational selection and preparation:* selecting appropriate educational and vocational goals; preparing for entry occupations and training for the future.

4. *Preparing for social and civic responsibilities:* educating oneself broadly; developing intellectual skills and

interests; acquainting oneself with social needs and problems.

5. *Preparing for marriage and family life:* developing a close relationship with one person of the opposite sex; forming wholesome attitudes toward marriage and having children; learning home management and child rearing.
6. *Building a philosophy of life and harmonious system of values:* accepting moral values; identifying with a religion; finding meaning in life.

Early Adulthood. The tasks and adjustments of early adulthood center around marriage, establishing a home, getting set in an occupation, and assuming the obligations of an independent, responsible citizen. Table 4.3 lists some of these tasks.

Table 4.3. Tasks and Adjustment of Aulthood

Early Adulthood (ages 21-35)

1. *Courtship and mate selection:* premarital testing of roles and dependencies; compatibility; changing partners.
2. *Marriage:* adjusting emotionally to partner; assuming new roles and dependencies; changing roles with parents; assuming new roles with in-laws.
3. *Family living:* tasks and roles in childrearing; managing finances' house-keeping responsibilities (cooking, chores, maintenance).
4. *Non-marriage:* adjusting to single life; developing new kinds of dependences.
5. *Occupational development:* completing training; problems of advancement and relocation; problem of combining work with marriage (women).

6. *Community responsibilities:* assuming roles m civic, public welfare, and charitable groups (PTA, scouts, fund drives, etc.); voting and political activity.
7. *Health:* maintaining health and vigor (weight control, exercise, keeping fit); adjusting to emotional stresses of bereavement, illness, or handicap.
8. *Social and recreational activities:* finding congenial social relationships in church, clubs, and activity groups; developing hobbies and interests; finding enjoyable recreations.
9. *Values:* continued formation of philosophy of life and religious meanings; self-actualization.

Middle Adulthood (ages 36-60)

1. *Marriage:* changing relationship to each other; decline of sex activity; changing roles with parents.
2. *Non-marriage:* continuation of adjustment problems from earlier adult years.
3. *Family living:* changing roles toward children; continuing financial and housekeeping responsibilities (now paying for children's education); assistance to grown children and their families.
4. *Occupational development:* "Peaking" in work; problems of retraining due to technological changes; reentry into job market (women); continued problems of advancement and relocation; preparing for retirement.
5. *Community responsibilities:* continuation of early adult tasks, but with increasing roles of leadership.
6. *Social and recreational:* continued, with increased importance as family disperses.
7. *Health:* of increasing concern due to aging.

Initially there is great uncertainty about selecting a mate and establishing a secure and intimate relationship. If the young adult has been successful in his earlier relationships with his parents, siblings, peer group, and others, his chances for a happy marriage are good. Children from broken or unhappy homes are less likely to have successful marriages. The young man looks to his wife for the MIR he found with his mother or other significant persons he has known, and the wife looks to her husband to play roles once played by her father or former boyfriends. Marriage involves the blending of whole cultures or families, not just the matching of persons alone. The marriage of two persons represents the interaction of significant systems based upon religion, nationality, and social and economic stratifications.

Many quarrels, especially in the earlier stages of marriage and courtship, represent tests of the strength of new roles and relationships. It is sometimes difficult for the new couples to relinquish their former dependencies upon parents. Later stresses are more often related to conflicts of roles and role expectations of the wife and husband. The young wife, for example, may be expected to be a good cook, housekeeper, nurse, lover, career diplomat, "mother," partner in sports, scout leader, teacher, and so on. As children are added to the family even further role changes are demanded. Now the husband must share the attention of his wife with others, and is expected to maintain the love and respect of them, at the same time he is expected to be prime disciplinarian.

The early years on the occupational ladder are tense ones for the young husband. How to complete the training he needs and how to meet the round-the-clock competition for advancement, while burdened by home responsibilities and financial obligations at a time when his income is low, puts quite a strain upon him. Added to such frustrations are the military uncertainties.

Somehow, many young couples weather the strains and come to assume contributing roles in their communities. There are others who cannot, and whose lives become torn by divorce, chronic debt, alcoholism, and meaningless drudgery in their work. The early adult years may be more stressful or more enjoyable than any other period of one's life, depending upon how well or how poorly a person has mastered the tasks and adjustments of his earlier years.

Middle Adult Years. As one moves into the middle adult years, many of the self-conflicts and doubts become intensified. Now the brutal facts are known, and hopes are waning for becoming the kind of person one might have been; time begins to run out. Both the man and the woman must live with their mistakes. Automation has increased problems of work dislocation and unemployment for the older workers. Women who reenter the job market after their families are more self-sufficient must accept unchallenging positions unless they have special training.

The woman who has concluded her childbearing may sense the loss of being needed as a mother as her children leave the "nest." She has come to depend upon their MIRs, just as they have come to depend upon hers. Then there are the adjustments required by aging parents. Former roles of dependency become roles of care and protectiveness for the young couple toward their parents. Friction may develop, too, because aging grandparents resent their dependency upon their children, and may try to maintain their roles of authority.

●●

5

Group Behaviour and Human Experimentation

The word "group" has many meanings. As we have already seen, one of the chief defects of social sciences is the use of terms in daily usage in scientific work. A number of controversies have arisen because the word, "group" has been understood in different ways by different people. In a broad way, there is the more general meaning of the word group and the narrower and the more specific meaning of the word group. We use the word when we are referring to any aggregation or congregation of human beings. When we see a number of individuals moving on the road in the evening in a city we call it a group. Similarly the people who are sitting in a bus or in a railway compartment are called groups. The pilgrims who are walking towards a town or a city to participate in a religious festival will also be called a group. On the other hand the members of a caste, the members of a club, the members of a state are also called groups. Since we use the same word to refer to various kinds of groups there is bound to be misunderstanding. We also use the term group when we are dealing with mere classifications. When we conduct a sample survey and interview a number of people of varying age, sex, education and socio-economic status we speak of them as groups. Actually these people have no relationship whatever with each other. They may never come together but still the social scientist when he makes an analysis calls them groups. Thus the term group is used in the following three kinds of contexts: *(a)* Where a number of persons are sitting together or walking together; here the essential thing is the physical proximity of a number of people coming together at a given time with or without any common purpose. *(b)* We also use the term group for mere classifications when we speak of tax-evaders, black-marketeers and so on as groups. These people may have no relationship with each other but because they have a common characteristic we put them together and classify them as a group. *(c)* Finally, we use the term group for the members of an organization with a definite structure, people who have a loyalty and a sense of belonging to the given group. In this part we will be dealing only with the first and the third kinds of groups because they involve social interaction. They affect the behaviour of the individuals in groups in more or less specific ways.

Concept of the "Group Mind"

Before we proceed further it would be useful to refer briefly to an old controversy regarding the concept of 'group mind'. Long ago the French sociologist Lebon used the concept of group mind to explain the various characteristics of crowd behaviour. He asserted that when individuals come together as members of a crowd, they are in the grip of a 'collective mind' and consequently their behaviour is very different from their behaviour as individuals. Terms like the 'group mind', 'folk mind the 'mob mind', 'collective consciousness' and so on are used to explain why individuals behave in a peculiar way when they are members of a crowd. Lebon said that when a person is a member of a crowd his conscious personality disappears and it is his unconscious personality, moved by the collective mind, or the group mind, that is responsible for his behaviour.

On the other hand we have McDougall who uses the same term group mind as a concept to explain the behaviour of individuals as members of highly stable enduring groups like the army or the church. He used the concept of group mind in order to explain the behaviour of individuals in such highly organized well-integrated groups. Among the outstanding characteristics of such groups he spoke of the continuity of the group, the self consciousness of the group with respect to its nature, composition, functions etc.; interaction of the group with the other groups with different ideals, purposes and traditions and the organization within the group involving differentiation and specialisation of functions.

Thus, the same concept of a group mind is made use of to explain the peculiarities of behaviour of the individual in a mob as well as in a highly organized group like in the army. Similarly, attempts have been made to explain peculiarities in behaviour of other groups like racial groups and caste groups. Allport has exposed the fallacies in this kind of analysis of behaviour. The concept of the group mind, a mind over and above the minds of the individuals in that group, is not necessary to explain the behaviour of the individuals forming the group. We should then speak, not only of a British mind, but also of an English mind, or a Londoner's mind, in order to explain the differences in the social behaviour of these various groups. In fact in any given town or city we will find differences in the behaviour of people living in the different areas. So we will have to speak of the mind of each area in the town or the city or even of each road or part of a road in a city. Thus, the whole problem reduces itself into absurdity. There is no doubt that there are differences in the behaviour of individuals forming different groups. There is also no doubt that an individual behaves in a different way when he is a member of different groups. For example, the student behaves in a disciplined manner when he is in the classroom or in the library, but his behaviour is quite different on the playground or in the cafeteria or when he is in the hostel. But to seek to explain these differences with the assumption of a group mind hardly takes us anywhere because we will have to invent a number of group minds to explain specific types of behaviour, not only of a group of individuals, but of the same individual as a member of different groups. In the recent years this concept of group mind has been completely abandoned. We speak more in terms of "group behaviour" rather than in terms of group mind. We explain differences in behaviour in terms of differences in the situation, differences in status and roles, differences in social norms and so on. Group behaviour is not a mere sum of the behaviour of individuals in a group since it is dynamic. Each individual influences the other individuals and is influenced by the other individuals. A number of field studies as well as experiments have been conducted in the last few decades which are of immense value to the study of the behaviour of individuals in group situations of various kinds. In the following chapters brief summaries of the work done will be given and an attempt will be made to determine the characteristics of behaviour in the various kinds of groups. We can study the behaviour of individuals in these informal groups. We will also study some of the techniques that have been developed to study the behaviour of individuals in small groups, Next we can analyse the behaviour of individuals in well-organized formal groups. In view of the fact that there are factions in our villages as well as in the towns, we may try to study why these factions arise and how we can help the individuals to become members of more integrated groups and overcome the disastrous effects of intergroup conflicts. We can next study the behaviour of individuals, when they are members in an audience situation and when they are members in the unorganized crowds. Finally we can study the problem of leadership and try to show the way in which the leader affects the behaviour of individuals in the group and how, his own behaviour is affected by the responses of the members of the group.

Behaviour in Together Situation

As we have seen above Lebon drew attention to the fact that there are differences in behaviour of individuals when they become members of a group. Laboratory experiments were started long back to study the way in which being with other people affects various psychological processes like attention, association, memory and so on. Allport got the individuals to work by themselves in 'alone situation' and put 4 or 5 people around a table and got them to do the same work in this 'together situation', each working independently. The time was constant for each person in the two situations. He made comparisons in terms of the quantity of work and the quality of work. He tried to eliminate the effect of competition by having all the subjects finish the work at the same time. He also prohibited any comparison or discussion of the results among themselves. He gave instruction that when they are in together situation they were not to compete with each other and that there will not be any comparison of the results to determine who has done more. Allport reported that the majority of the subjects showed improvement in speed and quantity of work in togetherness situation. He called them "social increments". He further found that the increase was greater for simpler tasks than for more complicated intellectual tasks. He found that the slower individuals improved their speed more than those who worked fast. As regards reasoning tests Allport reported that there was a lowering in the quality of work in the together situation even though the quantity increased on the whole. Finally, he found that in judging odours individuals avoided extreme judgments of pleasantness and unpleasantness when they were in the together situation.

The other results of Allport particularly regarding social increments have not been completely substantiated by other investigators. Dashiell repeated Allport's work. He got the subjects to work simultaneously in separate rooms and to do the same work independently in the together situation sitting around a table. He did not find consistent increases in speed in the together situation. This was further checked up by, modifying the conditions a little. Each individual was usually made to work alone at different times. It was found that when an individual worked alone his work was not as much as when several individuals worked simultaneously in separate rooms and when several individuals worked simultaneously at the same table in the presence of each other. This experiment conclusively showed that there was competition at a conceptual level when individuals were working in separate rooms at the same time. Physical presence of the other individuals is not necessary to bring about the element of competition. He also got the individuals to compete with each other in the together situation and he found that there was a big difference in speed when people competed with each other in the together situation in comparison with their working without competition in the together situation. In another set of observations he got two people to watch the man when he was working alone. Here also he found that there was a difference in speed between working alone when others are observing him work, and working alone by himself.

As we have already seen prestige suggestion operates on the work as well as judgment of individuals. Moore demonstrated that mere presenting of the majority opinion to an individual is effective in changing his opinion even though the other individuals are not present. As we have seen, the political propagandist tries to influence the judgment as well as behaviour of individuals by asserting that most people think that way or act in that way. Saadi and Farnsworth found that subjects tended to accept dogmatic statements very readily when they were attributed to people whom they liked. On the other hand there was resistance when the dogmatic statements were attributed to a disliked person. Thus, in our work, as well as in our judgments, we are affected by the opinion of the majority, opinion of the expert, the opinion of the people whom we like and so on.

Asch used a number of variables in order to study the influence of other individuals in together situations. He varied the size of the majority, the number of individuals disagreeing with the majority, and the structure of the task. The subject had to judge the length of lines by matching a given line to one of the three lines recorded on a card. The experimentor had planted 8 people who unanimously made an error. It was found that the naive subject was greatly affected when he found that the majority in the group differed from him. When an unstructured situation was given so that the differences between the stimuli did not provide an objective basis for discrimination, the subjects were influenced significantly by the erroneous judgments of the 'planted' majority. On the other hand when there were clear cut differences in the stimuli and when the majority made obvious errors the subject was not influenced. This experiment shows that the majority opinion was accepted when the stimulus situation was unstructured. On the other hand, when the stimulus situation was clear, there was no tendency to accept the majority opinion, though of course, the individual felt bewildered as to who was right, he or the majority. This is the typical situation when an individual in a group disagrees with some of the superstitious beliefs and rituals. Asch also tried to study the influence of the size of the majority on the judgment of the lines of the subjects. He found that when one person was planted the subject accepted the errors communicated by the other subject a few times. It increased to 12.8% when there were two people to mislead him. It further increased to 28.5% when there were three people to mislead him. It was found that further increments from 4 to is people did not lead to further increase in errors. Thus the size of the majority is an influence only up to a point. Beyond that point, increases in numbers, does not affect. The same thing also happens in life. Though advertisers are fond of speaking of "millions" actually when we see that the product is used by two or three of the people whom we know or whom we respect then we may change to that brand.

Barenda repeated Asch's work with children from 7-10 years of age. It was found that the children tended to err in the majority direction more frequently than the adults. It was also found that the children were not so much upset by the absurd and obvious errors of the planted subjects as the adults were. This indicates that the child perceives his relation to the other people somewhat differently from the adults.

Thus, when other people are observing or doing the same task the effect on an individual's experience and behaviour varies depending upon the factors, which are operating at the given time.

Uniformities of Behaviour

Bales and his co-workers at the laboratory of social relations at Harvard University have tries to study the problems of communication when small groups of persons ranging from 2-10 people attempt at problem-solving and decision-making. They have attempted to develop standardised methods of observing, recording and analysing data of the processes of interaction and communication. "One of our basic assumptions is that there are certain conditions which are present to an important degree not only in special kinds of groups doing special kinds of problems, but which are more or less inherent in the nature of the process of interaction or communication itself, whenever or wherever it takes place". The subjects are asked to consider themselves as members of the staff of an organization which has been asked to consider the effects of a case of human relations in a factory and advise the authorities as to why the people involved in the case behave in that particular way and advise them about what they should do. Each subject is given a summary of the case material. After each person has read it, the summaries are collected by, the experimenter. Thus, actually when they are discussing the case, different kinds of problems of communication arise. Firstly, there is the problem of information. Though the individuals possess facts relevant to the situation, there is some degree of ignorance and uncertainty, because they have just read only the summary of the case. They do not have all the facts, and they have not made a detailed study of the problem. It is only through the process of interaction they will be in a position to arrive at a definition of the whole problem and understanding of the whole situation. Secondly, there are problems of evaluation. Different members will possess somewhat different values to start with. They are asked to advise the authority about the course of action. In arriving at a common value judgment it is necessary for the group to arrive at this only through interaction. Thirdly, there will be the problem of control. Directly or indirectly each member will try to influence the other members in arriving at a group decision. The members face a number of alternative decisions or solutions, some satisfactory and some unsatisfactory. Consequently they will be influencing each other.

The group has accepted to perform a task and it has also accepted to reach a decision. In order to achieve this the individuals in the group have problems of communication as well as organization, which arise in the course of interaction. They have to communicate information as well as opinions and suggestions to each other. They have also to maintain social-emotional organization. Various kinds of solutions occur. These are checked up in terms of opinions that they have, and as a result of discussion, the individuals converge into some sort of a satisfactory solution. Thus there is a good deal of give and take as the individuals of the group discuss this problem and arrive at a solution.

Bales tried to categorise the behaviour of the individuals as they discuss the problem and arrive at a solution. He found that the whole process of interaction could be categorised into twelve categories and that the problem areas could be defined into the following four areas: positive reactions, attempted answers, questions, negative reactions.

Thus, the individuals may show solidarity or antagonism, tension release or intensification, agreement or disagreement, give suggestions or ask for them, give opinions and evaluation or ask for them, and finally give information and clarification or ask for them. Thus, in the process of communication there will be instrumental-adaptive task areas where information, suggestion and evaluations of these are made or sought. But when a group discusses a problem it is not a pure intellectual matter, it is also a social-emotional problem. We may like or dislike the information, suggestion and evaluations; more than this, we may like or dislike the individuals who give the information, make the suggestions and offer evaluations. Thus there may be either a positive or a negative attitude. The observers will be sitting in another room and observe through a one way mirror so that they are not seen by the participants. Besides categorisation of the behaviour they also note down who speaks to whom. In a general way Bales and his co-workers have found that in any small group of 2 to 8 participants 25% will be positive reactions 11 % negative reactions of the social-emotional area and 57% giving information etc., and 7% of asking information etc. Thus, two thirds of the behaviour in such situation pertain to question-answer area while the other one third pertains to the positive or negative social-emotional areas. It was also found that while there was some difference between the successful groups and the unsuccessful groups, the difference was not very much.

Bales and his co-workers also found that "groups with no designated leader generally tend to have more equal participation than groups with designated leaders of higher status". It was also found that in any group generally there will be one individual who does about 40 to 50% of the talking. When the size of the group is more than five it was found that in general about 3 people will do 80% of the talking while the rest of the people will do only 20%. Thus the man who is able to recall the relevant information, give a number of alternative suggestions and make evaluations of the information and suggestions of his own as well as of the other people, such a man emerges in the course of interaction as the prominent person of the group. This leads to the formation and differentiation of status. "Efforts to solve problems of orientation, evaluation and control, as involved in the task, tend to lead to differentiation of the roles of the participants, both as to the functions they perform and their gross amounts of participation.... Both qualitative and quantitative types of differentiation tend to carry status implications which may threaten or disturb the existing order or balance of status relations among the members". Thus in addition to thinking and reacting to the task on hand the members of the group have also the problems of their social and emotional relationships to solve. When there are personal anxieties or antagonisms, then the basic solidarity of the group is impaired. These emotional problems arise more and more when the problems of evaluation and control become more prominent in the interaction. The social-emotional problems lead to a kind of status struggle and this may lead to an increase in the rate of negative reactions. It was also found that when the status struggles are satisfactorily solved, then behaviour of categories 1, 2, and 3 will rise to the peak. There will be, not only a good deal of agreement, but there will also be an increase in the jokes and laughter reactions.

Togetherness Situation of Group Situation

As we have seen the togetherness situation has very elementary social properties. It is just the presence of other individuals, whether they are observing us or doing the same tasks. We do not have any group properties here. All the various individuals do not feel themselves as members of a group. For example, when various individuals like the pensioners, go to the bank or the treasury on the last day of the month to collect their pension, they are just in the together situation. But if they meet each other month after month they may gradually form into a group. They may have interactions with each other, which may lead to some kind of social norms and possibly also some kind of status relationships. It is possible that when a new man joins, the other members may have a different attitude towards him and probably after a few meetings, the new man may be assimilated and may feel that he belongs to that group. As Sherif writes, "In the course of repeated interaction over a time span among individuals with common motives or problems, togetherness situations become group situation. The appearance of a group is marked by the formation of structure (organization) and a set of norms. As individuals become group members in this process, differential effects of interaction process become more pronounced and more predictable in direction and degree".

HUMAN EXPERIMENTATION

Human experimentation and research ethics evolved over time. Much of the time, the subjects of human experimentation are prisoners, slaves, family members, or the experimenter himself.

Barly Modern Science

Among the first documented human subject research experiments were vaccination trials in the 1700s. In these early trials, physicians used themselves or their family members as test subjects. Experiments on others were often conducted without informing the subjects of dangers associated with such experiments. A famous example of such research were the Edward Jenner experiments, in which he first tested smallpox vaccines on his son and neighbourhood children. In an instance of self-experimentation, Johann Jorg swallowed 17 drugs in various doses to record their properties. Conversely, the famous scientist Louis Pasteur "agonized over treating humans," though he was confident of previous results obtained through animal trials. He consented to treat a human only when he was convinced that the death of his first test subject, the child Joseph Meister, "appeared inevitable." (Rothman 1993)

Early Twentieth Century

In the 1900s, the progress of medicine began to accelerate, and the treatment of research subjects also changed. The concept of human rights emerged, and with it came discussions of various codes of ethics of scientific disciplines.

Walter Reed's well-known experiments to develop an inoculation for yellow fever led these advances. Reed's vaccine experiments were carefully scrutinized, however, unlike earlier trials. (Brady 1982)

Medical experimentation has also been performed on humans without informed consent, both covertly and under coercion. The pretext of medical experimentation has been used as a justification for some atrocities. From 1932 until the 1970s, in the United States, citizens were experimented upon in the Tuskegee Syphilis Study.

World War II

During the second World War, Nazi human experimentation occurred in Germany. At the war's conclusion, 23 Nazi doctors and scientists were tried for the murder of concentration camp inmates who were used as research subjects. Of the 23 professionals tried at Nuremberg, 15 were convicted. Seven of them were condemned to death by hanging and eight received prison sentences from 10 years to life. Eight professionals were acquitted. (Mitscherlich 1992)

The result of these proceedings was the Nuremberg Code. It includes the following guidlines, among others, for researchers:

Informed consent is essential.

Research should be based on prior animal work.

The risks should be justified by the anticipated benefits.

Research must be conducted by qualified scientists.

Physical and mental suffering must be avoided.

Research in which death or disabling injury is expected should not be conducted.

In Japan, Unit 731 experimented with prisoner vivisection, dismemberment and induced epidemics on a very large scale from the late 1920s onward.

Immunity was exchanged with America after the war for a tiny part of the results, so that in postwar Japan Ishii and others continued to hold honoured positions. The Soviets were blocked from most access to those responsible by the Americans, who coveted the Japanese results. The effects were lasting and China is still working to counteract the effects of buried pathogen caches.

In 1940 in the United States, four hundred prisoners in Chicago were infected with malaria to study the effects of new and experimental drugs for the disease. Beginning in 1942, mustard gas experiments were conducted on 4,000 United States servicemen in order to study the effects on the human nervous system. These tests concluded in 1945.

Declaration of Helsinki

In 1964, the World Medical Association developed a code of research ethics that came to be known as the Declaration of Helsinki. It was a reinterpretation of the

Nuremberg Code, with an eye to medical research with therapeutic intent. Subsequently, journal editors required that research be performed in accordance with the Declaration. This document set the stage for the implementation of the Institutional Review Board (IRB) process. (Shamoo & Irving 1993)

Beecher Article

In 1966, anaesthesiologist Dr. Henry K. Beecher wrote an article, "Ethics and Clinical Research", describing 22 examples of research studies with controversial ethics that had been conducted by reputable researchers and published in major journals. Beecher wrote, "Medicine is sound, and most progress is soundly attained ... ". He believed, however, that if unethical research were not prohibited it would "do great harm to medicine". Beecher provided estimates of the number of unethical studies and concluded "unethical or questionably ethical procedures are not uncommon". (Beecher 1996)

Belmont Principles

The Public Health Service Syphilis Study was among the most influential in shaping public perceptions of research involving human subjects. When the press exposed the study, US Congress appointed a panel that determined that the PHS Syphilis Study should be stopped immediately and that oversight of human research was inadequate. The Panel recommended that Federal regulations be designed and implemented to protect human research subjects in the future. Subsequently, federal regulations were enacted, including the National Research Act, 45 Code of Federal Regulations 46, and 21 Code of Federal Regulations 50.

In 1974, the United States Congress authorized the formation of the National Commission for the Protection of Human Subjects in Biomedical and Behavioural Research, known to most people in research ethics as the National Commission. Congress charged the National Commission to identify the basic ethical principles that underlie the conduct of human research. To accomplish this task, the National Commission looked at writings and discussions that had taken place to date and asked, "What are the basic ethical principles that are used to judge the ethics of human subject research?"

In 1979, the National Commission met and published the Belmont Report. The Belmont Report is required reading for everyone involved in human subject research. The Belmont Report identifies three basic ethical principles that underlie all human subject research. These principles are commonly called the Belmont Principles. The Belmont Principles include respect for persons, beneficence, and justice.

After World II

United States: MKULTRA, Tuskegee syphilis experiment (The Public Health Service Syphilis Study), hepatitis experiments on children at Willowbrook State School, Jewish Chronic Disease Study (1963), San Antonio Contraception Study (1971), Tea Room Trade Study, Obedience to Authority Study (Milgram Study), dermatological experiments on prisoners at Holmesburg Prison in Philadelphia (see Hornblum 1998), and human radiation experiments.

United Kingdom: (voluntary) human experimentation at Porton Down in the 1950s, leading to the death of Ronald Maddison.

Pharmaceutical giant Pfizer came under fire in 2001 for allegedly testing meningitis drugs on African children.

The CIA ran an extensive toxicology and chemical! biological warfare programme in cooperation with the US military. The Edgewood Arsenal and US Army Medical Research Institute for Infectious Diseases at Fort Detrick in

Maryland were the main headquarters for such studies. The CIA developed many of the exotic toxins, incapacitants, mind-altering substances and carcinogens. The CIA attempted to use toxins to assasinate on Fidel Castro and other world leaders such as General Abdul Karim Quassim of Iraq and Patrick Lumumba of the Congo who nationalized Congo's mineral mines. Mind-control substances were studied to facilitate interrogation and toxins were used as weapons in assasination. One of the toxins that the CIA studied extensively was derived from red algae called dinoflagellate which produced the red tide. The MK-ULTRA project was a CIA run human experiment programme where prisoners and unwitting subjects were administered hallucigenic drugs in attempt to develop incapacitating substances and chemical mind control agents. The operation was run by Sidney Gottlieb. CIA researchers at the Bien Hua prison conducted human experiments on Vietnamese Communist subjects as a form of forture. The CIA was accused of murdering one of the unwitting subjects of the CIA LSD tests named Frank Olson. The CIA claimed that the man committed suicide as an after-effect of the LSD but the family exhumed the body and a forensic pathologist found evidence of blunt trauma to the head suggesting that he was knocked out before being thrown out of the window of a hotel. This supported the Olsen family's claims that CIA killed had him because he wanted to expose the MK-Ultra programme.

A Manhattan District Attorney opened up an investigation but was did not have enough evidence to prosecute. It is interesting to note that soon after this happened, the former CIA director William Colby committed suicide and he would have been a key witness. Dr. Sidney Gottlieb requested that William Colby destroy most of the documents on the MK-Ultra programme. Some suspect that the trial would produce information that would devastate the reputation of the William Colby and other officials involved in MK Ultra.

The CIA also recruited the University of Pennsylvania dermatology profesor named Dr. Albert Klingman to run the studies of the Holmsburg Prison. Kligman also conducted human studies on the effects of blistering agents on human skin. The US government also conducted numerous human radiation experiments.

Human Vivisection

Vivisection has long been practised on human beings. Herophilos, the "father of anatomy" and founder of the first medical school in Alexandria, was described by the church leader Tertullian as having vivisected at least 600 live prisoners. In recent times, the wartime programmes of Nazi Dr. Josef Mengele and the Japanese military (Unit 731 and Dr. Fukujiro Ishiyama at Kyushu Imperial University Hospital) conducted human vivisections on concentration camp prisoners in their respective countries during WWII. In response to these atrocities, the medical profession internationally adopted the Nuremberg Code as a code of ethics. This code of ethics does not prohibit vivisection on humans.

Human volunteers can consent to be subjects for invasive experiments which may involve, for example, the taking of tissue samples (biopsies), or other procedures which require surgery on the volunteer. These procedures must be approved by ethical review, and carried out in an approved manner that minimizes pain and long term health risks to the subject. Despite this, the term is generally recognized as pejorative: one would never refer to life-saving surgery, for example, as "vivisection." The use of the term vivisection when referring to procedures performed on humans almost always implies a lack of consent.

●●

6

Functional Development

In our discussions thus far it has been clearly evident that there is, in reality, no clear separation between the processes of change (development) in the physical organism and those, which refer specifically to the functioning of physical structures. This is to say, again, what has been emphasized earlier, that growth, maturation, and learning are in reality but three aspects of overall total change in the individual associated with his living through time. In this chapter we shall concern ourselves primarily with developmental changes in behaviour—with that aspect of total development which we call learning. The term "behaviour" is used here in a broad sense to include all activity, that is, all functioning of the whole person, vita1 overt motor, intellectual, emotional, and social. Learning is the primary aspect of development in all of these. For example, even though the level of a person's ultimate potentiality for intellectual development, as we shall see, is genetically determined and is therefore basically related to the maturation of physical structures, his specific mental abilities within the limits of that inherent potentiality are acquired (learned). He learns them as he copes with and gains control over, or accommodates himself to situations of everyday life. For example, a child's ability to "pass" a particular test of intelligence, or the abilities necessary for him to function effectively upon entering public school, presumably have or have not been acquired by him in his preschool experiences. The actual abilities to perform at any level of effectiveness in everyday life are products of learning.

Likewise, the child's original temperamental predispositions, constitutionally based as they are, also undergo change through learning. A very bright and active preschool child, for example, in one sort of home situation, with a particular set of parents, might become an effective and responsible social leader. This same child, had he been born into a very different family environment, could become a domineering, hostile, and emotionally disturbed person. The direction and extent of learning is a matter of great importance in relation to an individual's emotional adjustment and his total personality development. General adequacy of functioning in life is very largely a matter of learning in this broad sense.

Just as maturation refers to development as seen in body structure, so learning in a broad sense is developmental change in function ability that comes from functioning itself. Function is simply the doing of something. It is behaviour. It may be very simple, as in the case of a reflex act, or it may be very complex. It may be outwardly observable, as with motor behaviour, or it may be implicit and hidden from direct view, as with mental behaviour. Obviously, the very processes of living are processes of change. Living itself constitutes a continuous progression of changes in the way the individual (and his behaviour) interact with his environment. These are changes in functional facility due to the exercise of function itself.

Activity in all its forms is presumed to involve to some degree the functioning of three complex body systems: the sensory system, the nervous system, and the response system. This functioning, it is presumed, results in modifications of the physical structures as well as in their functioning. Functioning apparently leaves something in the nature of "traces" in the physical structure involved. Physiological psychologists are hard at Work trying to learn the nature of these traces. Some researchers are intensively studying what happens chemically at the synapses with the passage of nervous impulses through them as learning takes place (Deutsch and Deutsch, 1966). Others are concentrating on hypothetical changes in the molecular activity within the nerve cells involved as changes in function take place. In any event, as functional patterns are repeated and as new related acts are performed, portions of the already present traces apparently become strengthened and new traces are left. Complex systems of interconnected traces thus become established.

This means that the child's behavioural possibilities become augmented. His readiness for progressively higher-level functioning results not only from his, continuing biological development (maturation) but also from the actual functioning of the physical structures involved. This, of course, suggests the great importance of the kinds and amounts of stimulation and opportunity to function that the environment provides.

Obviously, there are various specific ways in which functional patterns become changed with experience. There may be an increase in the strength or the precision of an overt behaviour pattern itself. The golfer, for example, learns to combine an appropriate measure of strength with the accuracy of his stroke for each situation in which he finds himself. Or the change may be a matter of replacing a customary (natural) response with a different, more appropriate, and more effective one. It may also be a matter of becoming responsive to new and different stimuli in the situation. As we have seen, temperamental predispositions may be modified, strengthened or directed into different behavioural channels. Attitudes, values, a interests change through experience and through the interest of others.

Modification and augmentation of the more implicit "inner" forms of activity also come about through exercise and experience. These functional changes constitute cognitive development. Here, as in other sorts of functional development, structural "traces" also presumably take place, particularly in the brain. The psychological counterpart of this change in physical structure is developmental change in the "cognitive structure." In terms of cognitive theory, new learnings—new knowledge and new patterns of mental activity-are acquired only as they become incorporated or assimilated into this hierarchically organized cognitive system. In short, we learn new things in terms of what we already know.

What the child has already learned and how well he has learned to perform in relation to life's requirements and the expectations of his environment constitute his level of "ability." Ability means functional! effectiveness. Ability, in the sense of one's repertoire of available learned reactions to the environment and the tasks it presents, is what the usual tests of so-called intelligence purport to sample and thus measure directly. This repertoire of acquired responses, of course, depends upon one's capacity to acquire progressively more adequate behaviour patterns. It also depends upon the richness of the environment in terms of stimulation and opportunity to learn.

Factors Affecting Learning

Since the child is constantly doing something, he is constantly learning something. The conditions about him, the events that happen to him or about him, the attitudes and behaviours of others toward him, and many other aspects of his daily milieu all conspire to determine what he learns. But subjective, or "inner," factors also playa vital role as conditions of learning.

Many of the factors within the individual, which affect learning are discussed in other connections. The child's capacity to acquire new levels of functioning controls his learning at any particular point in his development. In other words, an individual must be ready to learn. This readiness level, as we noted above, results largely from both the biological maturation of the physical structure and the functioning of that structure.

Previous Learning

As mentioned earlier, a fundamental factor that determines not only that but how well or how rapidly the child learns in a given situation is what he already knows." In the words of Gagne (1965):

> The child who is learning to tie his shoelaces does not begin this learning "from scratch"; he already knows how to hold the laces, how to loop one Over the other, how to tighten a loop and so on. The child who—learns to call the mailman "Mr. Wells" among does not begin without some prior capabilities: he already knows how to imitate the words "Mister" and "Wells," among other things. The theme is the same with more complex learning. The student who learns to multiply natural numbers has already acquired many capabilities, including adding and counting and recognizing numerals and drawing them with a pencil. The student who is learning how to write clear descriptive paragraphs already knows how to write sentences and to choose words. [p. 21]

The parent or the teacher who is concerned with providing a child with optimal conditions for further learning must obviously use as his guide the child's already acquired capabilities. The child can learn only what he is "ready" (able) to learn in terms of his level of maturation and what he already knows or has learned to do. He is not "ready" to learn trigonometry if he has not yet mastered the operations of simple arithmetic.

Motivation

The problem of motivation is an extremely complex and involved subject, far too complicated for adequate treatment here. Our purpose therefore is simply to touch upon a few broad aspects of motivation that seem particularly pertinent to the matter of increasing individual functional effectiveness. In connection with the discussion of the relation between the learner's present capabilities and new learnings, the factors of interest and challenge are of crucial importance. Clearly, there must be a "felt need," as desire on the child's part, to learn the new material or to acquire a new level of ability. If the new learning task does not have a basis for its achievement in the child's present capabilities, he will be perplexed, frustrated, and discouraged rather than motivated with a desire to achieve the new learning. On the other hand, new material must be new enough to be challenging, not boring.

Relevance

A somewhat related motivational factor is personal relevance. If the material to be learned, or the tasks to be engaged in, are to have value in promoting personal effectiveness, they must have real relevance to and be consistent with the child's cognitive-emotional structure, including his self-image. Lecky (1951) discussed the principle of self-consistency as follows:

> According to self-consistency the mind is a unit, an organized system of ideas. All of the ideas which being to the system must seem to be consistent with one another. The center or nucleus of the mind is the individual's idea or conception of himself. If a new idea seems to be consistent with the ideas already present in the system, and particularly with the individual's conception of himself, it is accepted and assimilated easily. If it seems to inconsistent, however, it meets with resistance and is likely to be rejected. This

resistance is a natural phenomenon, it is essential for the maintenance of individuality.

The nature of the child's "mind," in the sense in which it was used in the above quotation, of course, varies greatly among children. It is true that preadolescent boys generally have a strong need to regard themselves and to be regarded by others as "manly." Sex typing, by the time children reach the age of 8 or 9 years, has been very effective in establishing an image of manliness in the minds of boys, as well as a general idealized cultural stereotype of the attractive female in the minds of girls. But self-images vary widely from one subculture to another, from one occupational group to another, and from one social class to another. A boy in a lower-class family whose father, and perhaps other adult males who play prominent roles in his life, are engaged in work requiring physical strength, courage, and skill—for example, in the operation of heavy machines—gets a very different view of what is manly than does a boy in a middle or upper-class home whose father is engaged in a high-prestige profession or business. Likewise, what a girl develops as her image of the appropriate or desirable feminine role is determined very largely by the models she lives with and associates with but modified in various ways by what she sees on television or the motion picture screen or what she reads.

With such wide differences in experiential background, what is self-consistent for one child in the way of subject matter to be learned or school activities and interest might be anything but self-consistent for another. What is regarded as "sissyish" by one boy might be quite consistent with another boy's conception of manliness.

Also, in our rapidly changing society, the whole picture of sex typing is probably changing. Such factors as the women's liberation movement and the more common entrance of women into occupations and activities formerly regarded as men's will tend to break up sex-role stereotypes and to make for less clearly defined sex models for children to follow. Some signs of confusion of sex-role images seem even now to be in evidence. Motivation to learn, of course, is often stimulated by external conditions. Interest, for example, a subjective factor so vital to school learning, often depends to a large extent upon the nature of the material to be learned, the objects to be examined and manipulated, and other physical facilities and surroundings provided by the school. Certainly the little boy's concept of manliness has its origin in the models of manhood about him and the portrayals of manly acts he is exposed to through the mass media.

Environmental Stimulation

Our basic assumption has been that all development comes about through the interaction of organism and environment. Activity in response to environmental stimulation is an outcome of such interaction. Since changes in the level of one's functioning result directly from functioning, and since functioning (activity) is basically a matter of responding to the stimuli, internal and external, that impinge upon one's sense receptors, the vital importance of stimulation—its quality and appropriateness as well as its adequacy—becomes clearly apparent.

Only in recent years has the importance of stimulation to development in general become an area of special concern and active research by psychologists. The earlier assumption was that the development of mentality, during infancy and early childhood, particularly, along with the development of the organism in general, was purely a matter of maturation. This assumption was especially apparent in the prevailing theories of mental development during the first quarter of the present century.

The first real challenge to this point of view came during the 1930s with the publication of the research findings of the

Iowa Child Welfare Research Station concerning the influence of nursery school attendance upon children's tested IQ's. The main implication of these studies was the importance of the factor of variable environmental stimulation in mental development. The controversy among students of intelligence provoked by the Iowa findings finally led to the only tenable position regard· ing the factors of mental development; namely, the interaction point of view. Mental development, in the sense of the acquisition of new functional capabilities, results from the interaction between organism and environment. Thus, optimal learning, from the beginning, requires optimal stimulation.

In the recent past a great deal of research activity has been concerned with the importance of environmental stimulation upon learning achievement in infancy, particularly. In this exciting area much is being discovered about the capacity of the infant to learn that was unheard of a few years ago.

Early Institution Environment

Studies made during the early 1940s of the development of children who were cared for as infants in orphanages and other institutions provided what appeared to be substantial evidence of the importance of access to and full contact with the "culture" from the beginning of extra-uterine life. These studies Bowlby, 1940; Goldfarb, 1943, 1944, 1945; Spitz, 1945; Spitz and Wolf, 1946; Brodbeck and Irwin, 1946) generally revealed a seriously retarding effect of the institution environment upon 'psychological development. Spitz (1945), for example, found that the measured developmental quotients (DQ's) of an orphanage group dropped " from 130-140 at 2 months of age to an average of 76 after 4 months in the institution, while a control group of home-reared infants with the same range of initial DQ's maintained the level through the 4-month period.

Spitz's studies, along with the other contemporary investigations of the effects of institution care, have been severely criticized on various counts, including the lack of rigorous control of variables. Pinneau, for example, noted specifically that Spitz's data did not support the conclusions and interpretations he placed upon them. Dennis and Najarian (1957) conducted a study in Lebanon which in certain respects was comparable to the Spitz (1945) study. These experimenters concluded that their data with respect to behavioural development in the institution environment showed no effect during the first 2 months of life but there was marked retardation during the age period of 3-12 months.

The newborn infant has very limited—if existing—visual ability to perceive objects and persons as such. He also almost completely lacks responsiveness to social stimuli. These observations are pertinent in relation to the Spitz study as well as to that of Dennis and Najarian. On the basis of such observations it seems reasonable to conclude that with normal infant handling and adequate physical care, and with the usual tactual and kinesthetic stimulation that babies receive, special kinds of stimuli from the external environment are relatively unimportant to functional development during the early days of life.

But developmental change in infancy is very rapid. As the baby's sensori-motor and perceptual abilities emerge and develop, and as (his affiliative need arises, the external environment—particularly its social aspects—becomes increasingly important to his continued cognitive and affective development. Dennis concluded that the relatively unstimulating environment of the foundling home, or the Creche of Lebanon, provided no opportunity for the infants to learn to perform the tasks or react effectively to the situations presented to them in the developmental test. Hence, their poor showing and low DQ's during infancy beyond age 3 months.

Adequate and appropriate stimulation is essential to optimal development from the very beginning. Many of the more recent observers of infant behaviour have stressed the importance to the newborn, especially, of the stimulation he gets from maternal handling and holding, particularly from close skin-to-skin contact (Brody, 1956; Escalona, 1953; Frank, 1957; Montagu, 1953; Wolff, 1959). Mirsky (in Escalona, 1953), for example, wrote: "In the earliest stages, an infant's security is a matter of skin contact and of the kinesthetic sensations of being held and supported". Frank (1957) views the "tactile-cutaneous processes" as an extremely important communication facility, particularly for the young infant. "The skin is the outer boundary, the envelope which contains the human organism and provides its earliest and most elemental mode of communication". Frank further stated:

> babies and children especially require these [tactual] contacts to recover from acute disturbances. Prolonged deprivation of such tactual contacts and soothings may establish in the baby persistent emotional or affective responses to the world, since his initial biological reactions to threat have not been allayed and hence may become chronic.

In discussions of the importance of tactile and kinesthetic stimulation during infancy little mention is made of their specific relations to learning or cognitive development. However, the theoretical question remains of whether or to what extent stimulation during childhood is a factor affecting the maturation of the bodily structure underlying functional capacity or whether the factor is merely a matter of the relative abundance of learning opportunities, as Dennis prefers to believe. The bulk of the evidence suggests that under conditions of extreme stimulus deprivation some irreversible retardation in development may result, but there is no doubt that, in a general sense, stimulation is a prime condition for learning as well as for other aspects of development.

There has also been some discussion in the psychological literature as to whether extensive intellectual stimulation during the early years of childhood might have deleterious effects, such as frustration resulting in negative attitudes, learning inhibitions, and overall psychosocial maladjustment. Fowler summarized the literature concerned with this question. He found much expression of opinion on the matter but little research evidence. On the basis of available evidence and in terms of his own experience in teaching a 2-year-old to read, however, Fowler came to the conclusion that the risk is minimal and that there is "promise of considerable success with other young children, given further refinement of techniques and method."

Cultural "Techniques" and Cognitive Development

In a provocative article published in the American Psychologist Bruner discusses another important aspect of outside influence, the culture, as an important interacting factor in cognitive development. He states:

> the development of human intellectual functioning from infancy to such perfection as it may reach is shaped by a series of technological advances in the use of the mind. Growth depends upon the mastery of techniques and cannot be understood without reference to such mastery. These technique are not, in the main, inventions of the individuals who are "growing up, they are rather skills transmitted with varying efficiency and success by the culture-language being a prime example.

Bruner goes on to point out that "techniques" such as language provide growing individuals with effective means of representing in a meaningful way "the recurrent features of the complex environments in which they live, Thinking, the major process of cognitive development (learning), is a process of selectively representing objects and events not immediately present to the senses and then examining them, categorizing them, recombining them, on otherwise manipulating them, and thus coming to know and better understand them. Thus the "culture" encompasses the exciting features of a changing, stimulating environment and at the same time provides the growing individual with the "techniques" for constructing models of the various features of the environment. "Cognitive growth, then, is in a major way from the outside in, as well as from the inside out".

Bruner then, taking his cue from an article by Washburn and Howell (1960), suggests the probable effects of such processes of individual development upon the evolutionary development of cognitive ability in mankind. He stated that "the principal change in man over a long period of years—perhaps 500,000—has been allopathic rather than autopathic. That is to say, he has linked himself with new external implementation systems"—systems that amplify human motor capacities (tools), that amplify sensory capacities (instruments of magnification, for example), and amplifiers of human capacity to think methodically and logically (language systems, theory explanation).

Any implement system to be effective, must produce an appropriate internal counterpart, an appropriate skill necessary for organizing sensori-motor acts, for organizing percepts, and for organizing our implement systems. These internal skills, represented genetically as capacities, are slowly selected in evolution. In the deepest sense, then, man can be described as a species that has become specialized by the use of technological implements. His selection and survival have depended upon a morphology and a set of capacities that could be linked with the allopathic devices that have made his later evolution possible. We move, perceive, and think in a fashion that depends upon techniques rather than upon wired—in arrangements in our nervous systems.

Nature of the Learning Processes

"Not only has man wanted to learn, but often his curiosity has impelled him to try to learn how he learns". To learn the "how" of learning is a difficult and involved problem. Human functioning is multifaceted, and since learning is functional change resulting from functioning, there are, in a sense, as many kinds of learning as there are kinds of activities in which human beings engage. Nevertheless, because of man's Curiosity about the fundamental nature of learning, certain individuals through the ages have speculated and experimented and thus developed explanatory ideas. During the past three hundroo years many of these speculations have been formulated into theoretical proposals designed to account for at least some of the facts of learning. Because of the extreme complexity of the problem, however, these theories generally leave much yet to be explained.

The various currently held learning theories can be grouped very roughly into two. main categories, each representing a fundamentally different point view in psychology. One is the so-called objective, or behaviouristic, point of view; the other is the relativistic, mentalistic, or cognitive point of view.

From the first point of view, the phenomena to be observed and measured are behaviour patterns, functions of the organism in response to stimuli of external or internal origin. This basic concept, the stimulus-response relationship, is symbolized by the simple formula S $\rightarrow$ R. From the mentalistic point of view, the basic phenomena of psychology are of quite a different order. They are conscious experiences, or responses to stimuli, rather than objectively observable acts. Ausubel (Anderson and Ausubel, 1965) draws a sharp contrast between these two viewpoints:

> Like the behaviouristic position from which it was derived, the neo-behaviouristic view focuses on publicly

observable responses and their environmental instigators and reinforcers as the proper objects of investigation in psychology. Consciousness is regarded as a "mentalistic" concept that is both highly resistive to scientific inquiry and not very pertinent to the real purposes of psychology as a science; it is considered an epiphenomenon that is important neither in its own right nor as a determinant of behaviour. Furthermore, say the neobehaviourists, it cannot be reliably (objectively) observed and is so extremely idiosyncratic as to render virtually impossible the kinds of categorization necessary for making scientific generalizations.

Exponents of the cognitive viewpoint, on the other hand, take precisely the opposite theoretical stance. Using perception as their model they regard differentiated and clearly articulated conscious experience (for example, knowing, meaning, understanding) as providing the most significant data for a science of psychology.

It should be emphasized that we are here comparing two very general classes of approaches and points of view about learning and that within each of these classes are different specific conceptions of the nature of the learning process. Hence the specific terms used by different theorists here roughly classed together are, only in a very broad sense, equivalent in meaning. Terms such as "behaviouristic," "neobehaviouristic," "objective," "associationistic," and "stimulus-response" characterize in a very general way a number of different specific points of view, but they have in common, however, a more empirical, mechanistic, and absolutistic way of viewing human behaviour. On the other hand, the second group tends to be more pragmatic and relativistic, being labeled by terms such as "mentalistic," "cognitive," and "Gestalt field," which are not univocal but are used by different students of learning with different shades of meaning. The term "mentalistic," for example, is more frequently used by adherents of the so-called objective school in referring to the contrasting theoretical position, whereas the term "cognitive" is most frequently used by the adherents of the position themselves in referring to their own viewpoint.

As a general rule, neither the neobehaviourists nor the cognitive psychologists are particularly concerned with the structure-function relationship, which is an integrating concept throughout this book. The adherents of the objective school are intent, rather, upon establishing cause-and-effect relations between objective happenings as symbolized by the S → R formula, whereas the cognitive people are concerned with the structure of consciousness and with the activities and manipulations within the cognitive field itself that constitute meaning and "coming to know."

Each in his own way however both the neobehaviourist and the mentalist, of necessity, deal with the individual-environment relation. As Bigge (1964) points out:

> The term interaction is commonly used in describing the person-environment process through which reality is perceived. Both families of psychology use the term but define it in sharply different ways. Whereas S→R association theorists mean the alternating reaction of organism, then of environment, Gestalt-field psychologists always imply that the interaction of a person and his environment are simultaneous and mutual-both mutually participate at the same time.

The objectivist is much more specific and atomistic in his references to the environment. He speaks of stimuli, and his objective is to identify direct connections between specific stimuli, or forms of energy, and discrete reactions. However, the present tendency is somewhat more often to use "stimulus situation" and "moral behaviour," thus recognizing the complexity of the behaviour-interaction process.

The cognitive-field psychologist, on the other hand, is more likely to see the environment in less specific and more relativistic terms. The cognitive field varies from moment to moment. It includes every aspect of the momentary situation—even the behaving individual himself—which, at the moment, plays a part in influencing behaviour. The individual, from this point of view, is not seen simply as one end of an acting and reacting bipolar system (S→R relationship) but rather as an integral aspect of the cognitive field.

Earlier Objective Theories

The problem in connection with learning theory is that of conceptualizing the essential nature of the changes-that is, the variations, the complications, the elaborations that take place in toe functioning person—largely as a consequence of his own activity. We shall first examine some of the more prominent theories of the objective school of thought.

"Trial and Error" Learning

E.L. Thorndike, one of the earliest American learning theorists, began publishing his work on learning (the associative processes) in animals in 1898. In Thorndike's thinking, learning is a matter of substituting effective or satisfying responses to stimulating situations for responses that do not produce satisfying results. For example, a hungry animal in a new situation will perform many ineffective acts in response to the total stimulating situation. Eventually it will, by chance, hit upon the act that will make food available to it. According to Thorndike's theory, the "bond" between the response that brought relief from hunger and the stimulus is strengthened. Learning has taken place. A "connection" is established between the situation and the response. The effective response from that time on, in that and similar situations, will tend to be substituted for the ineffective ones. This is trial-and-error learning, and Thorndike believed it to be the basic process in all learning (See Fig. 1).

It will be noted in Figure 1 that two sorts of stimuli are involved in trial-and-error learning: the internal stimulus (the hunger) and the stimuli, mainly visual, which arise from the various objective features of the situation and to which the animal has been responding differentially. The former, the internal stimulus of hunger, is referred to as the drive, the motivating factor that keeps the animal active until he chances to make the response that is instrumental in bringing satisfaction. Motivation brings about activity, which consists of random "seeking" responses to the various aspects of the situation and which is necessary if learning is to take place. Learning thus consists of forming connections (bonds) between specific stimuli and specific responses.

Thorndike's Laws of Learning

As an elaboration of his theory and based upon his observations and experimentation, Thorndike (1912) formulated his well-known laws of learning. Among his primary laws are the following:

1. **The Law of Exercise.** "Other things being equal, exercise strengthens the bond between stimulus and response". By the same principle, the lack of exercise of a connection tends to weaken it. This law, of course, in universally taken for granted and is implicit in all efforts of rote memorization.

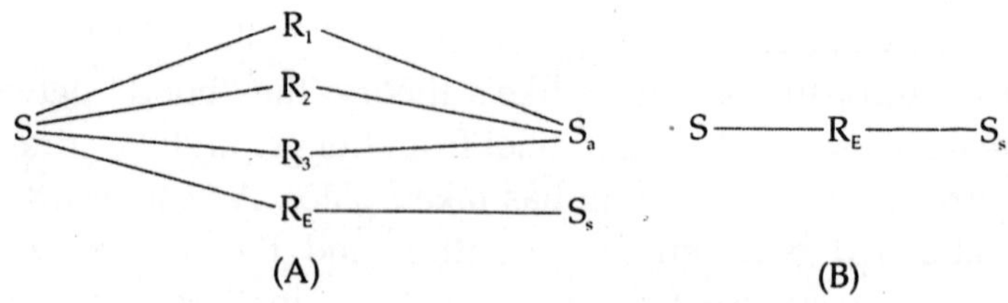

Fig. 1. "Trial and error" learning

A. A hungry cat is placed in a cage, but within sight and smell of food (S)—a strange situation for the cat. Being highly motivated it responds to the situation in a random manner-claws at the bars (R_1), tries to squeeze between the bars (R_z), bites at the confining walls (R_l). The result is a continuing "annoying state of affairs" (S_a).

In its random biting behaviour the cat chances to bite at the wooden latch of the door of the cage (R_E). The door opens and the cat gets to the food, thus bringing about a "satisfying state of affairs" (S_s).

B. With additional trials, random behaviour is eliminated. Only the successful response, R_E remains. A bond is thus formed between the situation, S, and biting the latch (R_E). The cat has learned a means of escape. (after Thorndike, 1924)

2. **The Law of Effect.** Again, in Thorndike's words, when a "modifiable" connection is being made "between an S and an R and being accompanied or followed by a satisfying state of affairs man responds, other things being equal, by an increase in the strength of that connection. To a connection similar, save that an annoying state of affairs goes with or follows it, man responds, other things being equal, by a decrease in the strength of the connection". This law has special significance in connection with current theoretical developments discussed later in this chapter.

3. **The Law of Readiness.** This law is concerned with the physiological functioning of the nervous system and its conductive units. Thorndike assumed that when an S→R connection is formed, a conduction unit, consisting of a specific set of neurons and their synapses, is established. These conduction units vary in degree of readiness to function according to the particular situation: "for a conduction unit ready to conduct, to do so is satisfying, and for it not to do so is annoying".

Thorndike and his followers in theory regarded this conceptualized process of establishing S→R bonds through the stamping-in effects of "drive reduction" and the stamping-out effects of unsatisfying or painful responses as the prototype of all learning.

Conditioned Response

From the work of Pavlov (1927), the famous Russian physiologist, came the concept of the conditioned reflex. A conditioned reflex is a unit of behaviour, which can be elicited by a previously inadequate (neutral) stimulus. The learning process in which this substitute stimulus becomes "connected" to the response is called conditioning. This process is as follows: the neutral stimulus is paired with an adequate stimulus—one that already has the capacity to elicit a particular response. In a series of joint occurrences of these two stimuli (the originally inadequate one slightly preceding in time the adequate one), the inadequate stimulus becomes functionally connected with (conditioned to) the response. The learning, in this case, is the acquisition of a substitute stimulus rather than a substitute response, as in Thorndike's formulation. Conditioned-response learning is diagramed in Figure 2.

Many experiments in conditioning have been conducted with both human and animal subjects, thus establishing it as an important learning process. Attempts have been made to expand the concept to explain all learning. Conditioned-response theory quite soon became a rival to the trial-and-error, Or connectionist, point of view that Thorndike developed.

Modifications and Extensions of Basic Concepts

Clearly, both trial-and-error learning and conditioning are importantly involved in functional change. Neither theory by itself, however, has been found adequate to account for all forms of learning. Each of them specifies conditions essential to certain forms of learning, but neither accounted for or explained the basic nature of the learning process. The quest continued, therefore, for more adequate theoretical models.

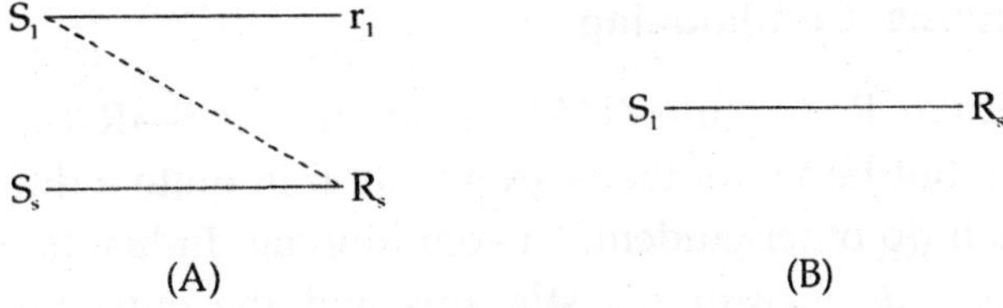

Figure 2. The conditioned response

A. S_s (nipple in Infant's mouth) Is the adequate stimulus to sucking and saliva secretion (R_s). S_1 (sight of the bottle), in the beginning, is a neutral stimulus to the feeding responses, eliciting perhaps only passive viewing behaviour. S_1 always precedes slightly in time S_1 (nipple in mouth). The two stimuli soon become associated in the infant's food-taking experience.

B. S_1 becomes a "conditioned" (adequate) stimulus to sucking, saliva flow, etc. (R_s).

All of the so-called associationistic, or connectionistic, theories and viewpoints that have been developed since the time of Thorndike and of Pavlov have taken as their starting point the concept of the stimulus-response relationship. They differ mainly with respect to the specific aspect of the process that is regarded as crucial in establishing the connection between the stimulus and the response. Some theorists have found it necessary to postulate certain conditions, or intervening variables, within the S→R sequence to account for the fixation of the bond; others have seen no such need.

Behaviour

John B. Watson, "the behaviourist" (1928), regarded Pavlov's simple formulation of the conditioned reflex as a sufficient basis for explaining all learning. He simply generalized and expanded the concept by adopting

Thorndike's law of associative shifting, which states that it is possible to "get any response of which a learner is capable associated with any situation to which he is sensitive".

Contiguous Conditioning

Edwin R. Guthrie (1952) is strictly an S→R learning theorist, but he views the S→R situation in quite a different way than do other students of conditioning. In his thinking, the objective, measurable stimulus and the outcome—the conditioned response or act that follows—are not the crucial elements in the associative, or learning, process. Rather, the elemental movements that constitute the response in the usual sense and the stimuli produce by or inherent in those movements become connected by association. This simultaneous contiguity between movement and movement-produced stimuli in a single occurrence, according to Guthrie, results in a conditioned response. Guthrie's one law of learning is: "A combination of stimuli which has accompanied a movement will on its recurrence tend to be followed by that movement". In an earlier paper Guthrie also stated that "a stimulus pattern gains its full associative strength on the occasion of its first pairing with a response". He thus saw no need for such concepts as reinforcement or drive reduction to explain learning.

Biological Adaptation Theory

One of the most influential theorists, particularly during the 1930s, was Clark L. Hull. More than any of his predecessors, Hull worked consistently toward a completely comprehensive formulation of the learning process. His theory is basically S→R conditioning. In contrast to Guthrie's principle of simultaneous contiguity, Hull assigned great importance to what happens during the brief interval between stimulus and response. He spelled out in considerable detail these happenings and the integrating elements involved, as he envisaged them, as "intervening variables," and he postulated a series of laws that defined the roles of these variables in the learning process.

Hull's (1943) formulation, however, is characterized particularly by his concept of reinforcement. He viewed a drive arising from a state of organic need as an indication that the conditions of survival for the organism are not being adequately met. When such a need, with its drive stimuli, develops, the organism becomes active, and when a particular act alleviates the need, that act tends to be stamped in as a biological adaptation of the organism. Thus, Hull theorized that all learning, whether of the conditioned-response order or the Thorndikian trial-and-error habit formation, might be explained in terms of a process of reinforcement from biological adaptation.

Concept of Reinforcement

The idea of the reinforcement following the response was first enunciated and studied by Thorndike (the law of effect, 1913). This idea has been elaborated and refined, however, by Skinner and his collaborators. The concept of the "feedback" is used to explain the operation of the reinforcer.

A common example of operant learning is seen in the behaviour of a young infant who awakens hungry from his nap. The condition of hunger motivates much overt activity. He kicks, squirms, thrashes about with his arms, sucks his fist (which finds its way into this mouth), and begins to cry lustily. As a result of his crying he is immediately fed. Crying, then, is the successful (instrumental) response bringing relief from hunger, which relief is a satisfying stimulus closely following the crying response. Crying when hungry is thus reinforsed, and the probabilities of the infant's crying when hungry thereafter are increased. He has learned a habit.

Figure 3 is a schematic representation of the operant-conditioning process.

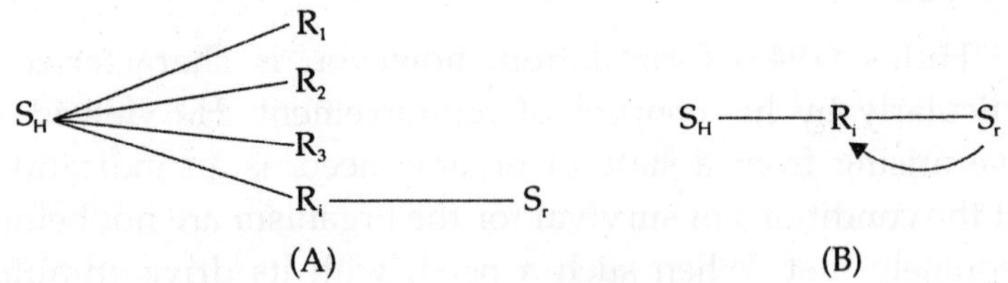

Fig. 3. Operant conditioning

A. S_H represents the stressful condition of hunger in the infant. He becomes restless, awakens, kicks, thrashes about, vocalizes, makes suckling movements, etc. (R_1, R_2, R_3, etc.) He then makes his discomfort known through lusty crying (R_i). Crying is the instrumental response which brings relief and satisfaction (S_r). He is taken up. Comforted, fed.

B. The satisfaction, SR_r, immediately following crying thus becomes a strong "reinforcer" for crying to bring relief from discomfort.

The reinforcing stimulus in the foregoing example was intrinsic in the situation. The natural outcome of drive reduction was the satisfaction of hunger. This was the primary reinforcer. There are, of course, other sorts of reinforcers, both positive and negative. In experimental work in learning, the experimenter can arbitrarily introduce various rewards for a correct response. Punishments which can be avoided only by the correct response are also in the nature of primary reinforcers because the rewards are made direct consequences of !he instrumental act. By their nature rewards are either pleasant and satisfying or unpleasant. An electric shock, for example, is naturally painful to the animal; a piece of candy or a colorful trinket is immediately pleasant to the child.

Secondary Reinforcement

Other stimuli, which are associated in time and are contiguous with the performance of the correct (instrumental) response and its primary reinforcement may quickly become conditioned (secondary) reinforcers. Such acquired reinforcers have become very important in learning theory as intervening variables. There are, of course, inherent in any learning situation extraneous stimuli that are potential conditioned reinforcers. Noises connected with the training box in animal experimentation may quickly acquire reinforcing value with respect to the instrumental response and thus become crucial factors in the acquisition and "shaping" of behaviour. Skinner (1959) found that conditioned reinforcers were important facilitating factors in the establishment of new behaviour patterns in pigeons.

In the course of a child's experience, as he interacts with his parents and others, he acquires many conditioned reinforcers that are of extreme importance in shaping his social behaviour and his personal development. The sight of his mother and the sounds she makes are paired with natural reinforcers to the responses he is learning. As a young infant he clings to his mother as part of the total pattern of nursing behaviour, which is instrumental to his experience of satisfaction and comfort (primary reinforcer). As he nurses he gazes into his mother's face and hears her voice. He also experiences olfactory, tactile, and kinesthetic stimulation, all of which become part and parcel of his experience of comfort and satisfaction. They thus become conditioned reinforcers to his clinging "attachment" to his mother. During the first 12 to 15 months of the child's life this attachment to his mother is an important factor in his emotional development.

Research in Operant Conditioning

Operant conditioning concepts and procedures lend themselves readily to precise mechanical regulation. Through early experimentation, largely with animal subjects, the basic principles of operant conditioning as formulated by Skinner in 1938, and others, have been established. As a consequence much interest has been generated in this approach to the study of learning. Recent work has more frequently involved human subjects. Many neo- behaviouristically inclined students of psychology have been attracted by the apparent ease with which the experimental variables and procedures involved in operant work can be placed under rigid mechanical control.

Currently the operant-conditioning approach is being used in the experimental modification of behaviour in a wide variety of human subjects. It is particularly adaptable to learning experiments with infants and young children because with them stimulus controls and procedures other than those built into the experimental apparatus are very difficult to impose. Hence, a greatly increased interest in the learning capabilities of infants and very young children has recently developed (Bijou, 1957; Brackbill, 1958; Rheingold, Gewirtz, and Ross, 1959; Simmons and Lipsitt, 1961; J. S. Watson, 1966, 1967; Kagan, 1963, 1964, 1972).

Exceptional children are also being studied by use of operant-learning techniques. Some of these studies have been concerned with the experimental evaluation of variations in stimulus controls (Bijou and Orlando, 1965; Ellis, Barnett, and Pryer, 1960). D. M. Baer and others have applied these approaches to the treatment of child problems such as thumb sucking.

In general, wide interest has developed in the operant-conditioning point of view, particularly in its potential application to practical problems. Skinner (1938, 1953, 1957, 1958, 1959) has led out in demonstrating as well as proposing practical applications of his procedures. He believes that "programmed" instruction, with reinforcements spaced properly to make them contingent upon the desired behaviour in the schoolroom setting, could result in greatly increased effectiveness in teaching.

D. M. Baer and Bijou and Baer (1963) particularly have applied operant principles to problems of social learning and the modification of social behaviour patterns in children. They have developed both laboratory (experimental) techniques and field-experimental methods of behaviour modification through the control of stimulus consequences. Baer, by means of a mechanized, talking puppet, demonstrated the use of attention and approval as social reinforcers. He also found wide individual differences in the effectiveness of these two classes of social stimuli as reinforcers.

Bijou and Baer (1963) even more strongly emphasized the importance of individuality among children at the preschool level and the different kinds of social reinforcements, which have, in the children's past, operated in the development of their social behaviour patterns and problems. These authors point out that through the careful and continuous observation of a child with a particular behaviour problem it is possible to note the various stimulus situations in which the objectionable behaviour occurs, and particularly its stimulus consequence in each instance. Then by controlling and modifying the consequences (reinforcements) of the behaviour, the behaviour itself controlled and modified. The most effective consequences which reinforce these undesirable patterns, they found, were the reactions of others, such as their attention, their approval or disapproval, their support and affection. Bijou and Baer briefly describe an example:

> This program has been applied to a case in which the child crawled rather than walked almost all of every morning. It was observed that the teachers responded to crawling with a great deal of attention and support, with a view toward improving the child's "security." When this was stopped, and the reinforcement shifted to the child's "upright" behaviour, the crawling weakened greatly within a few days and was largely

> replaced by upright behaviours of standing, walking, running, etc. Reversal of the contingencies to the old pattern reversed this outcome; reinstatement of the new contingencies again produced the new, desirable pattern. The result generalized well to the child's home environment, where the parents used similar contingencies (demonstrated for them in the nursery school) to maintain the upright behaviour.

Bijou and Baer also found that other behaviour problems such as excessive crying, over-dependence, aggression, and inattention could be successfully treated through this method of social reinforcement control. A lively interest is also developing in verbal behaviour as an area of operant-learning research (Rheingold, Gewirtz, and Ross, 1959; Skinner, 1957; Staats, 1961; Staats and Staats, 1959, 1962). This research area gives promise of considerable practical significance.

●●

7

Development of Maturity and Impacts

There is a tendency, of course, for us to think of human development as taking place during the so-called developmental period, that is, from birth, or conception, to adulthood. It is obvious, however, that significant changes occur throughout life and that many of these changes, particularly changes in functioning and in functional capacity, which occur after adult status is reached are of particular interest and importance in developmental psychology. Adulthood, which for convenience of discussion we divide into a number of sub-periods or phases, is our main interest and concern in this chapter.

As was noted in relation to the preadult periods of development, the lines dividing a particular postadolescent period or phase from the one before or the one following are very "fluid" and indefinite, especially as they are experienced by the individual himself. One tends to experience adult life especially as continuous and generally without demarking change points. As the course of development is studied, however, there are noted certain gradually developing differences in people, in physical appearance, in ready energy and vigor, in tenor and speed of performance, in social interests and behavior patterns, in predominant concerns and values, and in general circumstances of living, to name a few, which characterize, in a very general way, the various phases of adult development. We should be reminded also that with increasing age individual differences within a particular age group or developmental period become greater and qualitatively more diversified. People, as we have seen, differ widely in general developmental pace and in overall pattern. Hence any description of a phase or level of development can apply only in a very general senseto people in general and on the average.

The postadolescent, or adult, period of life, viewed as a whole and in general, covers most of the span of individual life-some 50-60 years. It is the time to which youth looks forward, often with some impatience, and the time to which the aging are wont to cling. It is also the time when inter-stimulation and the reciprocal influence of two generations, parents and offspring, are of exceedingly great importance in human development. For our purpose we shall arbitrarily regard the period of adulthood as consisting of four sub-periods: young adulthood, maturity-the peak years, middle age, and the period of retirement and aging.

Problems and Developments Tasks

This long period of adulthood obviously is not different from the preadult periods with respect to encountering and coping with problems of living and progressing through the various levels of development. Even with the great diversity of adult individuality and the tremendous range of changing circumstances and conditions of living there are certain common problems and crises to be met, certain developmental tasks more or less common to each general level or phase of adulthood. There is also a wide range of effectiveness and of degree of success with which these crises are met and these "tasks" are accomplished. Birren (1964) observed that it is rather difficult to avoid giving attention to the problems and crises of various periods of life. Thus the life span may appear, from technical descriptions, to be filled with a collection of unpleasant hurdles, resulting in a stumbling, bruising journey for the individual. This is obviously not a balanced picture, since there are many problems that are successfully met and solved and satisfactions gained that are not readily apparent. In a sense, the rewards of adult life take care of themselves; it is only the problems that demand attention. Because of rising standards, family life is often thought to have degenerated from a more blissful, earlier period of history; evidence suggests rather that the individual and the family have advanced not only economically, but also psychologically, mentally, and socially.

Young Adulthood

Typically the person at 20 is at his height in physical vigor. Intellectually he is likewise at his full capacity. For the several years preceding, he has been functioning at this full level of capacity "to acquire new modes of adjustment"—to profit from experience, to learn. With that level of cognitive facility he has become a very "knowledgeable person." He has established himself as an individual with his own thought patterns, attitudes, and feelings about people and society. By now he sees himself rather clearly in relation to the world and his functional role in it.

The young woman has had time to realistically assess her abilities and interests, and she is likely to have resolved for herself the marriage-eareer problem. She is likely already to have married or to have marriage as an immediate prospect. She sees herself either as combining marriage with career or as devoting her energies completely to one or the other and performing the social roles related to it. To the young man, however, it is most frequently not a question of "either/or." He must launch himself into an occupation or profession with the objective of supporting a developing family. He typically marries soon out of college and begins immediately an occupational career and the establishment of his family. This is the middle-class mode and ideal.

At this level of adult development, however, as we have already noted, there is extremely wide variation. There is much deviation from the usual or typical. Many college graduates who are professionally oriented are not yet ready to function in their chosen life's work. They must enter graduate or professional school for more years of study and preparation. These individuals often marry, and with the cooperation of their mates begin a rather difficult but relatively brief period of "subsistence living." But of course a great many youth sense no need or desire for college education. Many drop out without completing high school. These individuals tend to marry early and often find themselves with the responsibility of marriage and parenthood unprepared personally as well as economically. On the other hand, many of these young people, although they assume the functions and responsibilities of adulthood earlier than the social norm, make good adjustments and become constructive and desirable citizens in their communities.

Marriage and Family Establishment

Even though some individuals by choice or for other reasons remain single and childless, marriage and parenthood is the norm for adult living in our society. This means that the greatest and most important function of adulthood is producing and rearing the next generation. Adulthood, then, is the time when two generations of human beings normally, and by necessity, live together and interact in close relationships. Family living is the "effective," the all-important, aspect of the "environment" as a factor in human development. It is in these intimate relationships that the influences upon each other of the adult and the new generation are of crucial importance. And since the continuing quality of these influences and these relations depends very largely upon their beginnings, the great responsibility of young adulthood, when parenthood usually begins, becomes clear and obvious

It is then that the human developmental environment is established and its quality determined.

When a pair of young adults, a man and a woman, have established a love relationship with each other they are generally ready developmentally and psychologically to assume this responsibility. Erikson (1964) characterizes this level of development in the following words:

> It must be an important evolutionary fact that man, over and above sexuality, develops a selectivity of love: I think it is the mutuality of mates and partners in a shared identity, for the mutual verification through an experience of finding oneself, as one loses oneself, in another.

It is out of this shared experience· of a chosen active love of mates—this "mutuality of devotion forever subduing the antagonisms inherent in divided function"—that the capacity and the readiness to care in the most complete sense for the helpless new life comes to them as a product of their mutual devotion.

Time of Mutual Needs Satisfaction

The plight of the newborn infant, and the concurrent "emergency" nature of the parents' situation when a baby comes into their lives, have been reviewed in earlier chapters. Our primary purpose at this point is to emphasize the importance of a peculiar relationship of interdependence and mutual influence that exists between parents and their newborn baby.

We have already noted the state of "incompleteness," the profound immaturity and helplessness, of the neonate. We have also reviewed the radical changes in his physiological functioning which "nature" brings about at the time of birth for his survival. We have noted the tremendous impact of new living conditions with the great increase in external stimulation, which he undergoes at birth. His "desperate" need for the comforting protection and the warm and nurturant care of a mother is obvious.

The biological, particularly the neurological, state of development of the infant at birth precludes any conscious awareness on his part of his plight or of his need. The evidence from much skilled technical observation however, strongly suggests that the newborn baby does "sense." deep disturbance or comfort and quiescence, depending upon the degree to which these fundamental needs are neglected or are met in the infant's care.

Margaret Ribble (1944), an early observer of infant adjustment, was more specific than other writers concerning the importance of maternal care of the neonate. She related "mothering" to three kinds of sensory stimulation—tactile, kinesthetic, and auditory, all of which are adequately supplied normally by the mother as she takes the babry up, cuddles and fondles him, feeds him, and "talks" to him.

Most of the literature concerning early infant care, which has been well reviewed by Yarrow (1961), was concerned particularly with "maternal deprivation" and with special reference to satisfying the infant's nutritional needs. The more recent studies, however, tend to bear out the earlier observations of Ribble that sensory stimulation is perhaps of even greater importance to certain aspects of the child's development. Psychoanalytic theory, as we have seen, stresses the importance of sucking and the gratification of hunger in the establishment of an emotional attachment between the infant and his mother. Research findings, however, do not support the assumption of the allimportance of the feeding relationship. Heinstein (1963), for example, in a study of different modes of infant feeding found no support for the psychoanalytic emphasis upon the feeding relationship. There was no evidence of the superiority of breast feeding over bottle feeding in the formation of a mother-infant attachment.

Another study in this same area by Reingold (1956) resulted in similar conclusions. She studied a group of institution babies, which she "mothered," in the sense of providing much nurturant stimulation, in comparison with a control group who were given ordinary institution care. Her nurtured group showed significantly greater "social responsiveness" than did the control group.

The work of Harlow and his associates, although they did not work directly with human infants, has, nevertheless, contributed greatly to the theory of infant care in relation to development. In studies published in 1949 Harlow presented convincing evidence of the great importance of tactile and kinesthetic stimulation in the early development of his primate subjects. His results gave special emphasis to the importance of tactile stimulation, and one of his conclusions was that the emotional attachment between mother and child has little to do with the actual feeding process.

One important implication of these studies is that perhaps the most crucial aspects of nurturant care are not dependent upon the specifically maternal function of suckling the baby, that they can be equally adequately supplied by a fathering person. In other words, while in no way depreciating the importance of the care of the mother, the emphasis should be upon parental care. Perhaps our cultural stereotype of the male, and particularly of the father, as one without capabilities and therefore relatively free from responsibilities in relation to the care of his babies, needs revision. There is evidence that, from the standpoint of their mutual development, the relationship between a father and his children cannot be started too early in the children's lives.

Father's Role

The literature on child care indicates that our society in general is definitely mother-centered. This was the general finding in a survey of research literature by Nash (1965). One of his conclusions was that psychologists have adopted this cultural philosophy of child care [that in an industrial society the child caring function must be largely delegated to mothers], perhaps uncritically, and many appear to have assumed that it is both the only and the most desirable pattern of child care. In consequence, the majority of psychologists have not perceived the father as important in child-rearing, and this is reflected in their writings. Some psychologists have adopted the cultural assumption so thoroughly as to ignore the father entirely or even to deny him a position of significance.

This culturally determined concept of child care has further removed the father by enhancing the assumption that the rearing of children is a specifically feminine duty.

Sociologists, however, have more frequently noted and decried this culturally determined philosophy (Gorer, 1948; Elkin, 1946; Rohrer and Edmondson, 1960).

Although the writings of psychologists generally fit very well Nash's summary quoted above, at least one significant study has been published in which fathers' own feelings and conceptions of their roles in their families were investigated. The indication was that fathers tend not to see their roles as they have been portrayed by psychologists.

A study by Tasch (1952), although 20 years have elapsed since its publication, is especially significant in that she got her data directly from fathers, a fairly representative sample of urban American fathers living in the greater New York area. She interviewed 85 fathers of 160 children, 80 boys and 80 girls. The interviews were characterized as "flexible," which "allows for the emergence of a 'pattern of spontaneity and not a dutiful response'.

The results of this study showed that these 85 fathers felt quite differently and saw their role in their families as of greater importance than has usually been described by theoretical writers. These fathers did not see themselves as merely "vestigial" but rather as active participants serving a function in the care of the children, which they felt was in no way secondary to that of the mothers. As a group they felt that their role was not limited to "supporting" the family. The results generally indicated also that companionship with the children was highly valued by these fathers. The relatively small proportion of them who expressed dissatisfaction with their role usually gave lack of companionship with children as the main cause.

This is but one study of one small segment of our society. Obviously, the need is for more investigations of the father's role in the home in which other segments of the population are sampled. We need more reliable information from fathers themselves regarding the satisfactions they experience and the personal values they see in their relationships with their children. Evidence of a recent trend toward more attention given by researchers to the role of the father and his importance in child rearing is suggested in a study by Eron, Banta, Walder, and Laulight (1961). Their study was primarily concerned with parental child rearing practices in relation to aggression in children. The study involved ratings by fathers and mothers of their children's behavior in relation to their own interactions with the children. An interesting conclusion was that the fathers' ratings were more significantly related than were those of the mothers. They concluded: "Only recently studies emphasizing the importance of the father in socialization of the child have begun to appear. These results are further evidence of his importance both as the new method and the new dimension in childbearing research".

Caring as a Factor in Personal Development

"Care" has been described as the "virtue" and the need, which characterizes the young adult stage of development (Erikson, 1964). This is a time when babies are born, when the role and function of parenthood begins, and when the caring attitude is vitally important to the development and welfare of the new life. The exercise of this caring function is no less important to the personal development of parents, both father and mother. As Erikson puts it: "Once we have grasped this interlocking of the human life stages, we understand that adult man is so constituted as to need to be needed lest he suffer the mental deformation of self-absorption in which he becomes his own infant and pet".

The stereotype of the incompetent and uninterested young father in relation to the care of his infant was referred to above as a cultural creation perpetuated in the authoritative literature. We also reviewed some evidence that fathers generally do not see themselves as they have been described. It is undoubtedly true, nevertheless, and even though the "need to be needed" is equally strong in fathers as in mothers, that many young fathers have tacitly accepted the stereotype. They recognize the constitutional fitness of the mother and the priority of her relationship in the suckling and comforting of the infant, and with a sense of inadequacy and without encouragement they miss the opportunity and relinquish the privilege of early coming to know their baby through actually participating in its care. The young father sometimes does feel left out and even rejected by his wife in her preoccupation with her baby, and so he tends to become "his own infant and pet" in his own "self-absorption."

Harry Stack Sullivan (1953) stresses the great importance of the interpersonal interaction between parent and infant. He describes the "tension of need" in the hungry or otherwise distressed infant, the expression of which, in crying, induces tension in the parent. In his words:

> The tension called out in the mothering one by the manifest needs of the infant we call tenderness, and a generic group of tensions in the infant, the relief of

which requires cooperation by the person who acts in the mothering role, can be called need for tenderness....

Although Sullivan makes no specific mention of the father's role in this vital parent-infant interaction, the implication is clear that the father can function in many instances of "tension of need" in his baby by taking the role of the "mothering one." In such a relationship he can experience the "tension of tenderness." It is in this kind of parent-infant interaction that one finds and exercises his capacity to be completely other-centered, to give without demanding or expecting anything in return-a capacity that comes only with maturity.

Maturity, the Peak Years

People generally, by age 30, have passed through a period of settling down, and of becoming well launched and established in occupational and social roles which are likely to characterize them throughout life. Although men, particularly those who enter the professions, often marry after age 30, the mode is by then to be well into the period of family expansion with two to four children. Women generally marry at ages from two to five years younger than men.

Intellectually there has been development during the first 10 years of adulthood. By age 30, men and women have been functioning for some 15 years at the highest level of cognitive capaoity (Piaget, 1952; see Chapter 8). They have learned much and have profited greatly from experience. They have acquired many skills, mental, social, and occupational, and have grown greatly in wisdom. Their "crystallized intelligence" (Cattell, 1969) has grown.

For the majority of people this 15-20—year period of "maturity" is marked by serious and consistent striving for accomplishment. Occupational skills and effectiveness increase under conditions of business and occupational competition.

Earning power increases, keeping pace generally with increasing financial demands of the family. As the family grows in number the need for better and more adequate housing becomes acute. As the children grow their educational and cultural needs increase. In general this period of family development is a lively and challenging time of life.

Family Roles

As these changes in personal competencies, challenges, and conditions of living come about with family development there are, of course, accompanying changes in the complexity of family role patterns and relationships. The range of ages of family members generally widens considerably during this period of parental "maturity" as the parents grow older and as new babies join the family group. Thus role relationships increase in number and the potentialities for intergenerational influences multiply. In order to highlight some of these potentialities for personal development we shall examine in somewhat greater detail some of the typical family role patterns in Western society.

Adult Family Role Patterns

The man of the family finds himself functioning in a number of discrete roles as his family develops. The primary male role pattern would include functioning as an individual person, as a husband, as a home provider and "breadwinner," as a father, and as the male head of the family. The woman likewise assumes a parallel and complementary pattern of roles: as a person in her own right, as a wife, as a home maker, as a mother, and as the female family head. In actual family interaction, of course, roles and role relationships are much more varied in form and in affective quality than is suggested in these brief listings. It should be emphasized, too, that an individual functions in a particular role always in relation to someone else—another family member and that the quality, the meaningfulness, the effectiveness of role functioning and interaction vary greatly in different person-to-person combinations and from time to time and .in different situations.

All of these separate family roles and the relationships involved are important, encompassing as they do very largely the functions and the activities that constitute family life. However, since our main concern in this chapter is with relationships and interactions between the older and the younger generations of family members, the discussions that follow will deal mainly with those family roles which in a special way involve generational interactions with high potentials for stimulating personal growth, particularly the roles of being a person and the parental roles.

The Person

Functioning simply as an individual person may not immediately be thought of as a family role, but with careful consideration it may appear to be perhaps the most basically important area of functioning in family life. The degree of consistency and the level of autonomy with which the man of the family functions, for example, with his personal uniqueness and his pattern of idiocyncrasies is potentially of greatest influence upon the overall quality of family interaction and thus upon the adjustments and development of the younger family members. Individual uniqueness is always a factor in interpersonal relations. And as we have seen, the way a person "sees" himself and feels about himself largely characterizes him as a person. In a very real sense, one's "self-concept," his self-evaluations, his sense of personal worth and competence, are the easily read labels he pins upon himself, and these self-written labels are to a large degree accepted by others, even by members of his own family.

In another way a parent's sense of personal security is important. His or her ability, for example, to adjust with equanimity to such realities of life as growing older or approaching middle age—accepting without alarm or attempts to hide the loss of physical or athletic prowess or sexual potency or, in the case of the woman, the gradual loss of the bloom of youth or the onset of the menopause—is an important personal attribute affecting the quality of interaction with the younger generation.

Another exceedingly important factor in inter-generational relationships is a parent's ability as a person to listen—to value, to respect, to take into consideration, to integrate into his own thinking and expressions the views and opinions of those about him that may be at variance with his own. Nothing contributes more to a young person's sense of personal security and self-esteem than to have his ideas and points of view recognized and treated with respect.

Carl Rogers (1961) in his discussion of the characteristics of a "helping relationship" emphasizes the importance of making the person whom one wishes to help feel valued and accepted:

> Can I meet this other individual as a person who is in the process of becoming, or will I be bound by his past and my past? If, in my encounter with him, I am dealing with him as an immature child; an ignorant student, a neurotic personality, or a psychopath, each of these concepts of mine limits what he can be in the relationship. Martin Bubel', the existentialist philosopher of the University of Jerusalem, has a phrase, "Confirming the other," which has had meaning for me. He says "confirming means ... accepting the whole personality of the other.... I can recognize in him, know in him, the person he has been . . . created to become... I confirm him in myself, and then in him, in relation to this potentiality that... can now be developed, can evolve." If I accept the other person as something fixed, already diagnosed and classified, already shaped

by his part, then I am doing my part to confirm this limited hypothesis. If I accept him as a process of becoming, then I am doing what I can to confirm or make real his potentialities.

In the eyes of little children, of course, anything that Daddy does is good and to be emulated. But later, as adolescents, they often are embarrassed and annoyed by the personal and social habits and traits of their parents. To young persons in their quest for a sense of identity, the social and occupational status of their parents can be a matter of some Importance. It is important to them to be proud and not be ashamed of their parents.

The way ones sees himself and feels about himself can set the tone and quality of family life. The personal role one plays is significant also because of the basic relation between personality and one's functioning in the more specifically family roles.

Role of Father

As we have already noted, the rather generally accepted stereotype of the father of the family pictures his role as one of little consequence, particularly during the early periods of family establishment and child bearing. Little research has been published which concerns itself with how the father actually functions or with his attitudes and feelings concerning his participation in family activities and child care. The usual view of the father's role in the family during the more "mature" period of his occupational achievement and of family expansion (roughly ages 30 to 50) is one of even less family involvement. One often gets the picture of the suburban family matriarchy in which the father's participation in family management is limited to an occasional meting out of punishment for wrongdoing which the mother delegates to him. He leaves for work early in the morning and often returns late. Sometimes he is present for the family dinner but often not.

This description of parental functioning as children grow up applies particularly to the so-called affluent, middle-class American situation. However, according to the report of one of the committees of the recent White House Conference on Children (Report to the President, 1970), the pressures of living in our society which mitigate against adequate paternal functioning are no less strong for other segments of society (Bronfenbrenner 1971) The situation is indeed grim :

> for the family of poverty where the capacity for human response is crippled by hunger, cold, filths sickness and despair. No parent who spends his days in search of mental work, and his nights in keeping the rats from the crib can be expected to find time—let alone the heart-to engage in constructive activities with his children or to serve as a stable Source of love and discipline.

The whole tenor of the committee report was that the "pressures" of the social order, not the lack of parental concern, are to be blamed for the lack of constructive interaction between father and children.

The fact that fathers have to such an extent moved out of the lives of children, according to evidence, has led to a greater dependence on the part of children upon their age peers. This tendency has been shown in cross-cultural research Recently a comparative study was made of the strength of the influence of parents and of the peer group upon the activities in which teenagers engage (Condry and Simon, 1968). The results indicate that peer-group influence was significantly stronger, and this was particularly true of "antisocial activities." The implications are clear. The roots of alienation, drug abuse, and youthful delinquency generally largely reside in inadequate parental functioning, and especially the lack of father-child interaction in the home. Without parental influence and support the youngsters turn to their age peers. This weakness of parental influence appears not to be because fathers generally are not interested and do not desire to play a more adequate parental role, but rather because the pressures and conditions of modern having make it difficult, if not impossible, for them thus to function.

Role of Mother

Our emphasis upon the importance of the father's participation in family activities and the constructive influence he can wield upon the development of his children, is in no way to depreciate the basically important and essential role of mother. We have already noted earlier in this chapter and in other previous discussions how completely dependent the young infant is upon the nurturant care which his mother by inherent nature ordinarily is best equipped to tender him (Ribble, 1944; Spitz, 1965). Division of labor between the sexes, based as it is primarily upon biological sex differences, specially In our present highly individualized society, has made the care of Infants and children a special domain and responsibility of the female. And even though, as pointed out earlier, the human male is also capable of warmth and tenderness and of giving nurturant care to an infant, it is nevertheless true that these functions are inherent in and more natural to the role of mother. It is most frequently the privilege of the mother to participate with the infant in the mutual granting of the fulfillment of a basic need. As described by Montague (1966),

> The nursing couple, mother and infant, confer basic benefits upon one another—for when a baby is born a mother is or should be born. In the reciprocal relationship in which mother and child are involved it becomes increasingly evident that the gift they make to each other is their own selves-selves that are striving for fulfillment and development. The gift unaccompanied by the committal of the giver is arid, a

> mere thing unenriched by the human meaning to the recipient of the giver....
>
> The self grows by the interactive involvement with others: Whether the self develops as an affectionate one or as more or less defective m that quality, it will depend largely, if not entirely, upon whether its experience of such qualities from others, especially the mother, has been of a loving kind.

There is no doubt that the quality of mother's care and "interactive involvement" is an all-important factor in the child's personal development. As suggested, the role of mother and child rearer can be an exceedingly fulfilling one for a woman. Even in this age of independence for women and their freedom to take part in all aspects of modern living, of all the possible roles and activities open to them probably no other can be more satisfying and fulfilling than that of motherhood.

This is by no means to suggest that a mother should confine herself, and devote her life full time to the rearing and care of her children. This would not be good for either the child or the mother. Wise and adequate motherhood means "listening" to the individual child-being sensitive to his needs-and ordering the nature, the amount and intensity of his care in terms of his changing developmental needs. And this kind of mothering can best come from a mother who is enjoying the fulfillment of her own personal needs through participation in her chosen variety of interests, activities, and relationships outside as well as within the family. Work or other activity outside the home, under wisely ordered circumstances, can enhance the mother-child relationship.

However, as the children grow older and as the family expands during the period of maturity, the problems of child care and rearing obviously become much more complicated and the responsibilities often devolving upon the mother as the principal child rearer greatly increase. Typically the finds herself living a rather hectic life, managing the household and striving to meet the multifarious needs and demands of her children, often ranging in age from infancy into adolescence.

Social Influences and Parential Roles

But the child-rearing and training function has not always been, to the currently prevalent degree, the responsibility of the mother. The social orders of the past in relation to family management and child training and control 'have frequently been more patriarchal than matriarchal in nature. As we have seen, philosophies regarding child nature and development have varied within wide extremes. Children have been regarded as inherently bad, who must be purged of their evil tendencies through forced adherence to rigid rules and cruel punishment. At other times the philosophical viewpoint held that they were inherently good and that the environment the parents—must protect and encourage rather than coerce and punish. During the brief history of child development as a field of scientific study parents have received help and advice based upon research reports from child-care specialists and parent educators.

In a review of these recommendations through the years, however (Wolfenstein, 1953), one is struck with the wide swings in emphasis from one extreme position to another and the sweeping changes that have taken place in the nature of the advice and guidance offered. The need for parental guidance was keenly sensed, and as a consequence, advice was sometimes formulated from too little or incomplete evidence or evidence from separate facets rather than from a completely comprehensive view of an urgent problem or issue.

For example, at about the turn of the century the medical profession became especially concerned about the high infant mortality rate. Work in microbiology seemed clearly to indicate that unsanitary infant feeding was a main cause of this high mortality rate. As a corrective measure pediatricians in their guidance to parents instituted an infant-feeding regime emphasizing the importance of sterile conditions. It involved feeding from sterilized bottles and nipples rather than by breast. Specifically prescribed amounts of "formula" were to be fed to the baby according to a strict clock-dictated schedule.

The infant mortality rate was drastically reduced, but many of the leading students of infancy and infant care were unhappy with the regime because they felt that certain of the baby's important psychological needs were being ignored. Through their influence a. reaction to the strictly controlled regime began to develop. Certain influential child psychiatrists, pediatricians, and others (Aldrich, 1944; Trainham and Montgomery, 1946) began to recommend infant—care procedures that were a radical departure from the prevailing ones. Breast feeding, regulated in timing only by the demands of the infant, was strongly advised. Similar permissiveness was recommended in regard to training in elimination control. This new infantcare regime was accepted by many young mothers, but many others found it distasteful or difficult for various reasons.

In recent decades a more moderate point of view has developed as research findings have accumulated. It has been shown, and it is becoming more generally realized, that the feeling relationship between mother and infant-the quality of their interaction-rather than the specific procedures and feeding techniques, as such, are the important considerations.

As we have noted in other connections, the currently prevailing view concerning the determining factors of development, research, particularly at the infancy level, has been a major factor in bringing about this change in emphasis, with the recognition of the crucial importance of the role of mother in individual development. Since the mother in our highly technical industrial society is very often left with the complete responsibility of child rearing, the role of mother is one of awesome import.

To be sure, the mother's role changes with the growth of the child. The burdens of physical care obviously lessen as the child acquires "executive independence" (Ausubel, 1958). But there should be no change in the amount and quality-the genuineness-of love for the child as he traverses his course toward adulthood: from complete dependency and helplessness as a neonate to the status and autonomy of the 20-year-old, from the lovable and compliant 3-year-old to the loud and intrusive 11-year-old, from the "satellizing" and home-loving 6-year-old to the moody, rejecting adolescent. At all levels of development, particularly at times of stress or disequilibrium, when he may appear least to deserve it, the child needs especially the feeling that he is truly "understood" and loved.

Child-rearing Practices

The question as to how mothers actually perform their function of child rearing is one for which there is no single general answer. There are, of course, many factors of influence here. Every parent, obviously, has grown up in some kind of family situation and therefore is, herself or himself, a product of some kind of parental care and control. And there is observable in people generally a strong tendency in their own lives and in their relations with their own children to follow the family patterns of which they themselves are products.

Within our own society there are wide differences in ethnic background, and within each ethnic and cultural group there are sub-cultural and individual variations in childhood background experience. One factor which research has shown to be important in relation to child-rearing practices is the social class of the family.

Duvall (1962) described the social-class situation with respect to its self-perpetuating influences upon children:

> A child at birth is placed within society according to the social status of his parents. He may live out his life within the family status into which he was born. Or he may move up or down the social ladder as he "betters himself," or "is no credit to his family."...
>
> A family generally is aware of its status within the community. Others are seen as "our kind of folks," or "better than we are," or "not the kind of people we want to associate with," as families identify with those of similar status, those of higher and those of lower status respectively. Families who have social access to each other are generally considered to be of the same general social class.

People who thus feel akin to one another tend to think alike and to share common attitudes and patterns of living. Their somewhat common attitudes and beliefs about children and how they should be "brought up" tend to resist the influences for change from outside their social realm. Thus today, even though the popular magazines and other mass media make readily available to all levels of society the popularized versions of the most up-to-date research findings, as well as the current points of view of the "authorities" in the fields of child psychology, child care, and family relations, still rather wide social-class differences remain in degree of sophistication, in attitudes about children, and in patterns of child rearing.

As an environmental influence upon the development and adjustments of children, Rodman (1965) clearly portrayed the factor of social class:

> A person's social class background has a pervasive influence upon his life. The lower the class background

> the likelier they are to have tuberculosis, to die early, to receive poor justice in the courts, to bear illegitimate children, to have unstable marital relationships, to have "less" motivation and "lower" aspirations. An exceedingly wide variety of injustices and deprivations can be documented for members of the lower class of society-physical, social, occupational, legal, economic, and political. Even in affluent society like the United States, the poor are plentiful, and the injustices and deprivations they face are facts of life.

It is interesting to note how the relation between social class and childrearing practices changes as social change generally takes place.

Martha C. Ericson, in her 1946 study of child rearing in relation to social status, found some interesting differences between the middle and the lowerclass parents in their attitudes and expectations concerning their children:

> Middle-class families were generally found to be more exacting in their expectations. Training was generally begun earlier in middle-class than in lower-class families. In the middle-class families there was more emphasis on early responsibility, closer supervision of children's activities, and greater emphasis on individual achievement.

A more recent study by Sears, Maccoby, and Levin (1957) compared middle-class and working-class mothers as to their child-training practices. They found that the working-class mothers were significantly more severe in toilet training, more inclined to punish, scold, or shame the child for "accidents" than were the middle-class mothers. Although the working-class mothers did not begin this training earlier, they achieved results sooner. The difference in degree of severity and punitiveness was found in all areas of child guidance and training. The working-class parents tended to have less patience in relation to their children's dependency needs and to be less permissive with respect to such matters as sex play and expressions of aggression. Their method of control and training in these areas also tended to be more severe. In general, according to the findings of this study, the educational level of parents was the differentiating factor:

> The middle-class mothers, and those with higher education, seemed to impose fewer restrictions and demands upon their five-year-olds than did the working-class mothers of lesser education. Messiness at the table was somewhat less often permitted for children in the lower groups, and those children were subject to more stringent requirements about such things as hanging up their clothes, keeping their feet off the furniture and being quiet around the house.

In comparing the results of these two studies, M. C. Ericson's in 1946 and Sears *et al.* in 1957, some interesting changes seem to have taken place during that 11-year period. In the earlier study middle-class families were the "more exacting in their expectations" and seemed generally to be more inclined to pressure their children; they began toilet training earlier, and there was "greater emphasis on individual achievement." By comparison, the middle-class mothers in the later study seemed to be more relaxed and less severe and punitive than the working-class mothers.

The extent to which these apparent differences in middle-class attitudes and practices represent a real secular change in so short a period of time, of course, is not known. The group of Chicago mothers designated as middle class in 1946 mayor may not correspond in terms of socioeconomic status or cultural and educational level to a group of mothers from "two suburbs of a large metropolitan area of New England" 11 years later, also designated as middle class. The probabilities are great, in fact, that the two middle-class groups represented quite different positions in the social-class hierarchy with respect to the lower-class mothers and the working-class mothers with whom they were respectively compared. On the other hand, social change in general, including the great increase in amount of educational literature in the popular magazines during the II-year period, might well have brought about real attitudinal changes. Perhaps as middle-class mothers, as a group, became better educated and gained a fuller understanding of child nature, they lost some of their tendencies to pressure their children. Perhaps they did become more relaxed and permissive. It is likewise possible that parents at the lower socioeconomic levels, with generally improved economic conditions and more leisure, became more upward striving and thus more inclined to subject their children to "more stringent requirements" and to impose upon them more restrictions and demands in their efforts to obtain the "better things of life" for their families.

Home Environment

The crucial importance of the general home environment in the lives and the development of children has been noted in earlier discussions in this book. Undoubtedly, the influence of the female head of the family is by far the most important factor determining the character and quality of the home environment as she performs her roles as a person, as a wife, and particularly as a mother and child rearer. This was found to be an exceedingly important home-environmental factor in the child's cognitive development. Hess and Shipman (1965), referring to the restrictive sort of mother-child interaction, a mode which is generally devoid of individualized communication "specific to a particular situation, topic or person," made this brief summary statement:

> This environment produces a child who relates to authority, rather than to rationale, who although often compliant, is not reflective in his behavior, and for

whom the consequences of an act are largely considered in terms of immediate punishment or reward rather than future effects and long-range goals.

Parent behavior in relation to the development of children has been studied for many years. Champney (1939), at the Fels Research Institute, developed a set of rating scales for appraising parent behavior. Continued work with these scales, using factor-analytic methods, resulted in a suggested set of parent behavior "syndromes" or clusters. From his analysis Roff (1949) interpreted and labeled these parent attitude-behavior variables as follows:

I. Concern for child

II. Democratic guidance

III. Permissiveness

IV. Parent-child harmony

V. Sociability adjustment of parents

VI. Activeness of the home

VII. Non-readiness of suggestion

Some Family-Life Patterns

In an earlier study the reactions of teenage children to questions regarding their home situations were analyzed. Four quite distinct common patterns of family life were revealed (Stott, 1939). This study involved groups of youngsters, over 1800 in all, representing three different home settings: the urban setting, the small Midwestern town, and the farm in the open country. The questionnaire to which the youngsters responded included questions about the amount of recreation the family enjoyed both outside and in the home, expressions of affection in their relationships with their parents, the extent to which they confided in their parents, their feelings about the personal characteristics and habits-personal and social-of their parents, the health, both physical and emotional, of their parents, and the recency of parental punishment.

The responses of the three residence groups were analyzed separately by means of factor analysis. The outcome of the analysis, in each case, was a number of clusters of questionnaire items (factors), which were interpreted as family-life patterns, each distinct and different from the others. Each factor could also be interpreted as representing a particular type of family environment.

These common factors were *(1)* a pattern of intra-familial relationships which is characterized by "confidence, affection and companionability"; *(2)* a somewhat related, but less inclusive family-behavior pattern of "congeniality"; *(3)* a factor involving family discord-frequent punishment of the child and misconduct—which predisposes the adolescent child to offer criticism of his parents behavior; and *(4)* a "nervous tension" factor, "nervousness," or something which the child interprets as nervousness on the part of the parents, and usually such other items as illness of parents and infrequent demonstrations of affection. This investigation also showed significant relationships between family pattern and the personality adjustments of the children. For example, "self-reliance" in the sense of being independent in working through personal difficulties and problems was associated to a significant degree with the family pattern of "confidence, affection, and companionability" in all three residence groups. A measure of "personal adjustment" bore a similarly significant relation with that same family-life factor.

More recently E. S. Schaefer (1959, 1961) used factor analysis in his study particularly of maternal attitude and behavior material. The outcome of his analysis was the suggestion that the many variables of maternal behavior can be brought into the framework of two main factors. These factors, again, are regarded as dimensions, or continua. They were interpreted as love-hostility and control-autonomy. Figure 1 is based on a model developed by Schaefer, which summarized his findings.

Emotional Behaviour Dispositions in Parents

The two main axes in Schaefer's "circumplex" (love-hostility and control-autonomy) seem fairly well to fit Allport's (1960) concept of the "behavior disposition". As we noted, people do differ in their capacity to love, and if a random sample were rated on the continuum represented by Schaefer's lovy-hostility axis they might conceivably occupy positions ranging from the complete lack of love (hostility) to full capacity and disposition to love. Likewise, people by disposition vary in their relations with others from the need to control and dominate to the tendency to allow freedom and autonomy to others. Such dispositions in parents have much to do with the quality of the environment in which their children develop. In many studies and discussions of parent-child relationships through the years these two variables have been central.

Parental Love

Authoritative writers of the past 50 years have not always been in agreement concerning the matter of parental love. Perhaps in no other area in the study of child nature and parent-child relationships has professional thinking differed more sharply or changed more radically. Freud, in his earlier writings (1913), contended that too much "parental tenderness" "spoils" the child and causes difficulty for him in later life when he must be satisfied with lesser amounts of love. A few years later the American behaviorist-psychologist J. B. Watson much more strongly expressed the same point of view. In his book (1928), written specifically for parents, he devoted a full chapter to the dangers of too much mother love. Among his pronouncements were the following:

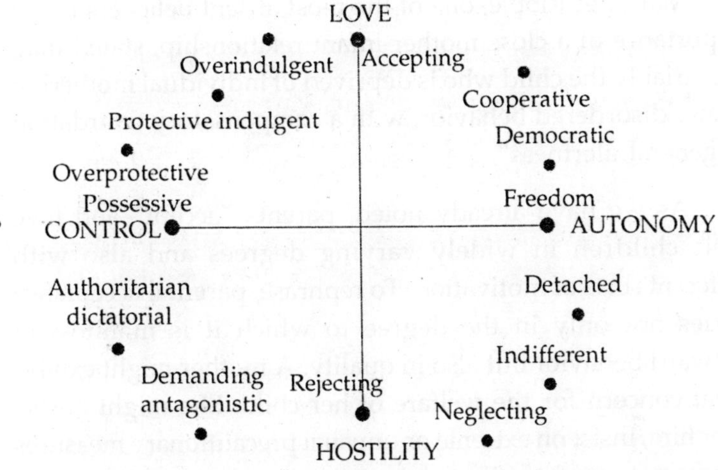

Fig. 1. E.S. Schaefer's (1959) circumplex model for maternal behaviour.

> All too soon the child gets shot through with t00 many of these love reactions. In addition, the child gets honeycombed with love responses for the nurse, for the father, and for any other constant attendant who fondles it. Love reactions soon dominate the child... In conclusion won't you then remember when you are tempted to pet your child that mother love is a dangerous instrument? An instrument which may inflict a never healing wound which may make infancy unhappy, adolescence a nightmare, an instrument which may wreck your adult son or daughter's vocational future and their chances for marital happiness.

During the decade or two following Watson's manual, the pendulum of expert opinion swung to the other extreme in relation to infant care. A number of writers have been equally forceful in stressing the child's need for parental love and its crucial importance in the emotional development of young children (Ribble, 1943; Spitz, 1945, 1946).

Margaret Ribble, one of the most ardent believers in the importance of a close mother-infant relationship, stated that, "invariably the child who is deprived of individual mothering shows disordered behavior, with a compensatory retardation in general alertness".

As we have already noted, parents "accept" and love their children in widely varying degrees and also with different sorts of motivation. To rephrase, parental acceptance varies not only in the degree to which it is manifest in outward behavior but also in quality. A mother might exhibit great concern for the welfare of her child. She might "fuss" over him, insist on extreme or unusual precautionary measures to insure his safety. She might worry a great deal about him when he is out of her sight and in other ways express in her overt behavior what appears to be great love for him, while in reality, and probably completely unconsciously, her basic feeling toward the child may be one of profound rejection. Her over acceptance, her exaggerated expressions of love and concern for him, may be unconscious compensation for or protection against her basic rejection and lack of genuine love for him.

Pseudo acceptance of a child in some instances may be exploitative due to emotional immaturity in the parent. The manifest "love" is not genuine love for the child but rather an exploitation of the child in unconscious self-seeking efforts on the parents part to obtain gratification of unmet emotional needs (I. D. Harris, 1959). The overt behavior of parents in relation to their children, in most instances, however, probably truly expresses, in varying degrees, genuine acceptance and feelings of love on the one hand or frank rejection on the other.

Individuals, as they become parents, do differ rather widely in the capacity and disposition to love and accept their children (Porter, 1955; Sloman, 1.948; von Maring, 1955). Evidence also indicates that this capacity is determined to a large extent by the quality of the parents' own family experiences as children.

Parental Domination

The other personality variable in parents, which perhaps has much to do with the quality of parent-child interaction, is the tendency, or need, to dominate. When this tendency is strong, it is rooted in the individual's own early childhood experiences and is likely to remain throughout life as a fundamental personality characteristic. In a person thus strongly disposed to be the dominant one, to be "boss," and always to be right are, to him, vital sources of inner security. To be domineering is not to be understanding of the other's feelings or opinions. As a parent, the dominating personality is likely not to be sensitive to his child's individual needs or to have much sympathetic understanding of childish idiosyncrasies, Parents vary widely with respect to this variable also, from complete domination and strict and rigid control to over-permissiveness and even complete relinquishment of control.

Figure 2 is a hypothetical representation of these two dimensions of parental disposition in relation to each other.[2] The area in each quadrant of this figure may be taken to represent the various possible combinations and degrees of that particular relationship. For example, over-solicitousness may be combined with domination. In the upper-left quadrant can be plotted any degree of each of these tendencies il1 relation to the other. Or over-solicitousness might be combined with over-permissiveness in various combinations of degrees as represented by the upper-right quadrant. The area in the figure, which might be enclosed by a circle at the center represents the most desirable positions on both continua, the positions that would be occupied by the "warmly dependable" and "understanding" mother as described by I. D. Harris (1959). These mothers would give their children abundantly adequate wholesome love, and they would neither rigidly dominate and control them nor be over-permissive, but would be understanding of the children's individual needs, giving them freedom within carefully defined limits and positive guidance as they come to need it.

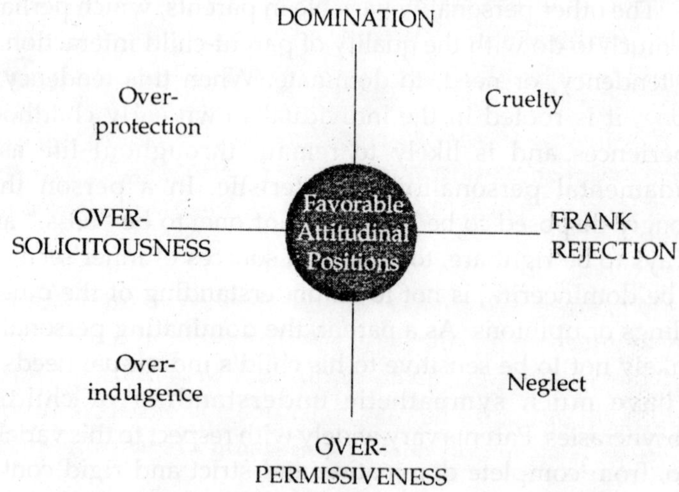

Fig. 2. Possible combinations of, and relationships between, the domination. permissiveness and the love-rejection dimensions of parental attitude with suggested outcomes in terms of treatment of the child.

As is indicated in Figure 2, too much domination combined with over-solicitousness might be expressed in overprotective behavior toward the child, but when strong domination is combined with rejection the most likely result is cruel treatment. On the other hand, over-permissiveness and the relinquishment of control combined with over-solicitousness would mean overindulgence, while the combination of no control with frank rejection would mean complete indifference and neglect.

There are many other ways in which these dispositions toward feeling and behavior might combine in the make-up of individual parents. A hovering, "smothering" mother for example, who compulsively overwhelms her child with "love," conceivably might be neither domineering nor over-permissive. It is likewise quite possible that a parent who cares not at all for her child and frankly shows her lack of feeling for him in her behavior might also occupy a middle ground on the domination versus over-permissiveness continuum.

These descriptions of the parent-child situation, of course, are oversimplifications. Normally there are two parents, each with a different pattern of attitudes and behavior dispositions in relation to others. They are not likely, therefore, to react to their children with the same quality and intensity of feeling. Both father and mother might love the child, but the quality of that love and its expression might be quite different. (One might be inclined to dominate while the other might be on the side of permissiveness.) The number and variety of interaction patterns that might develop within the triad-father, mother, and child-are indeed many. For a child to be overindulged or neglected, overprotected or treated cruelly by a mother or a father may be deeply disturbing to him. But in each case the effect might be softened and partially compensated for by more healthy and constructive relationships with the other parent or with other significant persons in his life.

Middle Age

During the extremely busy period of life, which we have labeled "maturity" time passes very rapidly. As one pauses to note that his older children are approaching adulthood, one almost suddenly realizes that he is getting along in years. He may find himself quite unprepared when he is confronted as if by a stranger instead of his little boy or girl. He is forced to realize that his family is beginning a new phase in its development and that he is not as young as he once was.

This period of middle age is generally a time of crisis. It is often a time of personal disturbance for parents, for example, as they face the fact that they are now among those who are called middle-aged. It is often hard for them to admit that they have already taken on the "middle-aged spread" with its associated limitations; that she must now accept herself as she is with the quieter charm of her age, and he must accept the fact that he is "slowing down," that he no longer can be the model of strength and dexterity for his son, who now can "run circles around him." Reuben Hill (1951) describes some of the difficulties of middle age:

> Both husband and wife face similar problems in the middle years because ours is a youth-oriented society. As a young country, America has rewarded youth and depreciated middle age and later maturity. Our schools and families have failed to prepare us systematically for the roles that may be played in middle age, and consequently we approach the period with great hesitancy and conflicting emotions. This is true whether we are burdened with the task of launching our children into jobs and marriage or are facing the complications of a childless old age. For it is in the middle years... that men must face their diminishing bargaining power in the market place, their decreasing vitality and powers of attraction for women, and their limited stature as persons in a society geared to poor-boy-makes good stories. It is in the middle years that the wife-and-mother discovers that raising one or two children and caring for a husband turns out not to be a permanently full-time job.

Evelyn Duvall (1957) presents another view of the problems of the modern middle-aged wife:

> Being middle aged today is a far more powerful business than it used to be in the days of Whistler's mother. Then, a woman in her forties was physically and psychologically ready to retire to her knitting. Today's middle-aged woman, thanks to better nutrition, medical services, lightened burdens at home, and shifting feminine roles, still has a "head of steam up" both physiologically and emotionally. She is apt to be vigorous, often feeling better than she has in her whole life. Even menopause, the dread of women in earlier eras, now can be taken in stride. The woman with nearly grown children has found her strengths and weaknesses.
>
> Now, suddenly, before she is quite prepared herself for it, her children are no longer children, they are taking their confidences and their loves outside. Her husband is engrossed in the peak of his business or career. Her house is in order. And she-where does she go from here? What love can take the place of those so suddenly torn away? What tasks can absorb the energies and the skills that cry for channeling? If she clings to her children she is a "Mom." If her interests in her husband's work becomes too absorbing, she is a "meddler." If she spends her time in a dizzy round of matinees, bridge parties, and beauty parlors, she is a "parasite." If she devotes herself to a quest for her soul through devious cults and sundry religions, she is suspect.

Perhaps with the changes in interests and values which seem to characterize youth generally today, women will enter the later stages of maturity having been less intensely involved in their motherhood role and be better prepared personally for the changes in role relations as the children mature. By the time the grown-up stage in family development is reached, however, family interaction patterns most certainly have changed in character and in the kind of meaning they have for the persons involved. In certain of these relationships the changes may constitute real crises in individual lives.

Middle-aged Husband-Wife Relationship

Even though the intimate relationship between husband and wife, as such, does not involve direct interaction between generations, it is none the less the most basically important of all family relationships. It has greatest potential for affecting the development, the welfare and lives of all family members. This is an especially important consideration at this point because a number of factors and conditions, both intraindividual and interpersonal, typically take place, which characterize this period of personal and family development.

We have already referred to the disturbing effects in both men and women of facing the facts of middle-aged life. There is research evidence indicating that many couples fail to find happiness together in their relationship during these middle and later years (Bossard and Boll, 1955). As we have seen, there are a number of factors, which contribute to this unhappy situation. For the wife, there is the fact that her function as a mother has markedly changed. It may seem as though she is jobless. Her children no longer need her mothering. Now, when she especially needs her husband's attention and closeness to fill the gap, she often becomes unhappy with the inadequacy of her middle-aged husband as a sexual partner. Then there is the problem of the menopause.

Husbands, most frequently in their fifties, according to this research (Bossard and Boll, 1955), are also often unhappy in their marriages but for different reasons. The data (information furnished by their brothers) identified two rather different groups of husbands:

> One consists of men who have attained some degree of prominence and success in their chosen field, only to

find that their wives have not kept pace with them m their upward climb. Such men often make a determined effort to remain loyal to their mates. Some of them are reported as succeeding; other fail and are aware of it; and still others appear to their siblings as failing in spite of outward evidence of success.

A second group of men, identified by their siblings, are in their fifties and have failed, absolutely or relatively, in their occupational efforts. Such failures lead them to rationalizations: they never had a chance to succeed they say. Their wives were no help to them. If it were not for the handicaps, which their wives imposed, they would have succeeded, as did other men. Such wives become scapegoats for the failure of their husbands.

The Menopause

The end of the reproductive period in women, the menopause, or climaoteric, usually comes during the fifth decade of a woman's life. The average age is 49 years. Physiologically it is brought about by a decrease in the hormones secreted by the ovaries.

The endocrine glands constitute a complex, interrelated, and marvelously timed and regulated system of chemical control in the organism. The cessation of the ovarian function, which ends the possibility of bearing children, is the result of a complex glandular operation. The pituitary gland, the master regulator of chemical balance, along with the thyroid and adrenal glands, is involved.

The cessation of menstruation, of course, is only one of the phenomena, which Occur at the climacteric. There is a complex interrelation between the glandular system and the nervous system, particularly the autonomic division. Not only are changes in glandular functions initiated by the nervous system, but it is in turn affected, with varying degrees of influence, by glandular secretions. The autonomic system is the nervous network, which controls the vital life processes of the organism. Since these processes have much to do with the emotions and with the sense of physical well-being-arising from respiratory and heartbeat rates, digestion, and other physiological processes—a train of subjective effects occur which can render the person uncomfortable and disturbed. Investigations during the past few decades, however, suggest that in the past the menopause has been used as a popular excuse for personal inadequacies in many instances. It has been estimated that only about 20 percent of women actually experience any serious menopause problems and that these difficulties are of short duration. It appears that the menopause, the dread of women in former times, is now "taken in stride" by modern women.

Relations between Generations

As we have already seen, the middle-years period generally is a time of emotional crisis and many factors and circumstances seem to conspire thus to characterize it. In many instances not only is there strain and tension in parents due to their personal problems, but it is also a time of serious intergenerational conflict. Sometimes the emotional upset of the menopause in mother coincides with the adolescent's quest for independence and his search for personal identity. In such instances each is so involved and preoccupied with his or her own subjective concerns as to be unable to respond many constructive way to the emotional needs of the other. In reaction the adolescent's tendency to alienation and rejection may be reinforced.

A recent research team has concluded that if a father is really to assist his boy through the problems and conflicts of adolescence he must himself achieve resolution of his own life problems peculiar to middle age and the declining years. This developmental task of maturity Erikson (1968) has called the crisis of "integrity." Integrity is the eighth and last of his "stages" in the life cycle of man, that is, in his development of personality.

> In the aging person who has taken care of things and people and has adapted himself to the triumphs and disappointments of being, by necessity, the originator of others and the generator of things and ideas—only m him the fruit of the seven stages gradually ripens. I know of no better word for it than integrity. Lacking a clear definition, I shall point to a few attributes of this stage of mind. It is the ego's accrued assurance of its proclivity for order and meaning—an emotional integration faithful to the image bearers of the past and ready to take, and eventually to renounce, leadership in the present. It is the acceptance of one's one and only life cycle and of the people who have become significant to it as something that had to be and that, by necessity, permitted of no substitutions. It thus means a new and different love of one's parents, free of the wish that they should have been different, and an acceptance of the fact that one's life is one's own responsibility. It is a sense of comradeship with men and women of distant times and of different pursuits who have created orders and objects and savings conveying human dignity and love. Although aware of the relativity of all the various life styles, which have given meaning to human striving, the possessor of integrity is ready to defend the dignity of his own life style against all physical and economic threats. For he knows that an individual life is the accidental coincidence of but one life style with but one segment of history, and that for him all human integrity stands and falls with the one style of integrity of which he partakes.

The alternative—the opposite—to this sense of integrity is "despair." This integrity crisis involves the increasing awareness of declining powers, loss of self-esteem and the like, with loss of motivation and the tendency to give up. With this attitude of despair on the father's part, according to the study referred to above, a distorted relationship with his adolescent son develops which leads to irrational conflict ("annihilating fights") in which the father defines the son as "an incorrigible criminal" instead of discussing specifically what he does and its rational implications.

By contrast the father who has successfully resolved his personal problems—who has achieved "integrity"—"gives his son an example of commitment to values and meanings" and even though the son may strongly disagree with those values and meanings their differences are dealt with in what the authors call a "loving fight." The son can then identify with his father's sense of commitment and react to him generally as one who is trustworthy.

Clearly, the middle years of family development are crucial years in the lives of both the older and the younger generations. This, of course, is the period of the much-discussed "generation gap." It is a time when a sense of implicit trust in the integrity of his parents is of crucial importance to the young person. Such a trust, as we noted earlier, has its origin during the months of early infancy and is developed throughout childhood in the interactions between the child and parents who are truth—worthy. With a background of early experience of always feeling that he was listened to and understood in trying moments and emotional upsets, now as an adolescent he is likely to express freely his impatience and intolerance with things as they are, his need to change the world, or to get away from it, knowing that he will be listened to with respect and understanding rather than with ridicule and condemnation. With such a realization he is much more disposed to listen to his parents and to recognize their more mature judgment.

Period of Retirement and Aging

During these later stages of human development people tend normally to continue to do what they have become accustomed to do. Subjectively, as an aging person, one's self-image tends not to change. Normally one's declining vigor and decreasing capacity for accomplishment come upon one in imperceptible degrees. Time alone, as viewed in relation to accomplishment, seems to pass with ever-increasing velocity. In this gradual process of maturing, as one brings together and orders one's life values, as one succeeds in achieving personal integrity (Erikson), or because of personality difficulties, when the crises of life have been so devastating that the person experiences only varying degrees of discouragement and despair-these processes of development also continue throughout the middle and declining years.

Actually, it is only as signal or crisis events take place in one's life-as a physical disability or a debilitating illness strikes, or as one suddenly realizes that his daily activity routine is about to be interrupted by "retirement" that one accepts the fact that he has reached a critical point and is entering a new epoch in his life.

Decline in Sensory Acuity

During these years of maturing, however, development generally is of the nature of gradual decrement, rather than increment in functional facility. "Generally with advancing age there is a reduction in sensory acuity as consequence of injury, disease, and changes in the structures and functions of neural tissues that are probably due to primary aging". Some of reduced acuity can be tolerated without an obvious effect upon behavior. Nature has provided a wide margin of safety in that there ordinarily is more sensory input than is necessary for the adequate detection of signals. However, with wide individual variation, there is, on the average, a gradual impairment of both visual and auditory acuity well underway by age 40. The frequencies of defects and impairment in these two vital sense modalities rise in positively accelerating curves. By age 60 the frequency of serious visual defects has reached approximately 70 in 1000 population.

Changes in the functioning of the interoceptors, or sense receptors in the internal organs, proprioceptors, or muscle sense receptors, and the vestibular receptors in the inner ear are usually not so apparent and are more likely to take place without any conscious awareness. But these sources of information are very important to appropriate movement in and orientation with respect to space. Without proprioceptive information, for example, one would have to guide his gross motor activities entirely by vision. Without these senses and without vestibular sensitivity one would be without the sense of balance. Gradual loss of acuity in these sense modalities generally comes with advancing age. Our knowledge of the extent and the rate of impairment of the various senses in relation to aging has been considerably expanded by recent research sponsored by the U.S. Department of Health, Education and Welfare. This information is reported in summary by J. E. Birren (1964). The extent to which these sensory losses become seriously disturbing factors in the life of the aging person obviously largely depend upon the nature of one's life's work and the kinds of activities in which he likes to engage. Serious impairment of vision, of course, is always debilitating. The loss of hearing especially has damaging effects upon one's ability to interact socially and intellectually with others and thus can seriously interfere with one's need to function in the ongoing life of the family.

Retirement

One of the inevitable consequences of aging is the necessity eventually to withdraw from one's major occupation or field of activity. The idea of retirement has widely differing connotations for individuals in the various occupations and vocations. Some persons, particularly those engaged in physically taxing kinds of work, as in industry, tend to anticipate with eagerness the time of retirement. They look forward to a time for rest and freedom from the tyranny of the time clock. For persons in the more sedentary types of work, like positions to the academic field, retirement is generally something to be delayed as long as possible and an event for which they are often unprepared when it comes. For the self-employed—the independent craftsman or the business or professional person—retirement is not something imposed upon him when he reaches some mandatory retirement age.

In many instances, however, retirement is the most abrupt and radical change experienced by an active individual, and it often involves rather serious adjustment problems. These difficulties may be largely family problems of adjusting to reduced financial income. They may be primarily personal in nature. In any event retirement for the person himself should not mean cessation of activity, a time with nothing to do. On the contrary, he should "retire" into a new and planned program of activity, a program designed by and for himself personally and for which he has thoughtfully prepared himself. Nothing can facilitate the aging process more or can contribute more to physical and psychological decline in a person who has always been active than to withdraw completely from active life.

Personality and Retirement

Growing old, and particularly withdrawing from the stimulation and pressures of the work-a-day world, like any other significant change in one's environment and one's pattern of daily activity, can bring about noticeable change in one's personality. The concept of "personality" is many faceted. The "core" of one's personality, one's "individuality," resists change. One continues to be the same person, uniquely himself throughout life. But just as truly, one is never a completely finished product" as long as one is alive and interacting with the world about him. Changes in outlook, in attitudes, ways of reacting to life situations do take place as one's effective environment changes.

As Birren (1968) suggests, "a working definition" of personality with reference to aging is that it is "the characteristic way in which an individual responds to the events of adult life". There is no doubt that an individual's response to life's events as he sees them from the perspective of retirement and the other accompaniments of aging are often different in certain respects than they were before as a more active participant in those events.

Research evidence indicates also that an individual's responses to the events associated with retirement and to the changes brought about in his daily routine depend very largely upon his basic personality structure, his unique pattern of basic dispositions to respond to life's events generally. A general conclusion of one extensive investigation of the adjustments to retirement of men at different ages was that those who were better equipped personally to stand up under stress generally adjusted most successfully to retirement (Reichard, Livson, and Peterson, 1962). These were the ones who found satisfaction in their activities and were able to accept change with greater equanimity. They were also those who tended to seek rather than to avoid social contacts.

These investigators, in the statistical analysis of their data, were able to describe three groups of men of roughly similar patterns of behavior dispositions who adjusted "successfully" to retirement. These groups were the "mature" ones described as having a more constructive approach to life, the "rocking chair type" who were inclined to be dependent upon others, and the "armored" type who maintained well-developed sets of psychological defenses against anxiety.

Thus it may be concluded that not only does one undergo some change in personality in the broad sense as a result of retirement, but the manner in which one adjusts to the associated events and the changes depends upon one's "individuality"—one's pattern of basic dispositions to respond to life in general. Women who have worked for years also have their problems of adjusting to retirement, adjustments which in most instances are probably not greatly different from those of men. Since it has been only in relatively recent years that great numbers of women have taken permanent jobs and have entered the professions, published research is lacking regarding their particular retirement problems.

Developmental Tasks of the Aging

The meaning of the fact of growing old, to the individual, and the manner in which he responds and adjusts to it, as we have seen, is largely a matter of basic personality—of the individual's pattern of emotional behavior dispositions.

> Some older men and women become petulant, demanding and difficult to please. Life becomes a burden for them, and for those who care for them. They resent the "insults of aging" as they gradually lose their physical attractiveness and powers, their jobs and status, their loved ones, and their former sources of satisfaction and fulfillment.
>
> Other aging men and women find the "golden years" of life the most fruitful of all, as they gather the harvest of a lifetime and keep on vigorously growing to the very last. Oliver Wendell Holmes observed in his later years that being seventy years young is £ar better than being forty years old.

However, even though one's basic personality structure may have become pretty well organized quite early in life and may, therefore, continue to play a dominant role in determining the nature of one's responses to his world of events, the many years of unique experience prior to the period of aging undoubtedly also have had much to do with his capacity to adjust "successfully" to the processes of aging and the events associated with them.

Modern society, of course, is concerned about the problems of retirement and aging and has provided facilities to help the individual prepare ahead of time for a comfortable and constructive old age. The elderly, it is assumed, can also be helped to realize that the period of aging is a period of development with its particular "developmental tasks" which can be worked at and achieved.

As has already been implied, perhaps the most basic general developmental task of the elderly person-one upon which much else depends-is that of finding life meaningful for himself. And it is in the areas of interpersonal, particularly intergenerational, relationships that the most meaningful and rewarding experiences lie. A number of specific developmental tasks for enhancing family interaction in the later stages of family development have been proposed.

●●

8

Behaviour and Social Interaction

Every social reaction starts with the perception of the other individuals or groups. Consequently the problem of perception is basic in social interaction. When you meet an acquaintance in the street and start talking to him the social interaction begins with your perception of that individual. If you had not perceived him there would not have been any social interaction whatever. Why did you see him? It is not just because he happened to be going. Several times as we may recall, some of our close friends have told us that we did not perceive them even though we passed by them. Probably they even tried to attract our attention and failed. It is not true that we perceive all the human beings whom we pass by. As Gardener Murphy has put it: "In the light of the strong trend in recent years to lift the problem of social psychology from the behaviour level to the level of awareness of social reality the need for reasonable theory of the process of social perceiving has become imperative". Particularly since the work of Rorschach and Murray has established that personality determines perception, we find that the problem of social perception is receiving considerable attention. What are the determiners of social perception?

Structural and Functional Factors

How do the laws of perception operate in social interaction? As long ago as in 1940 the present writer indicated "all perception whether illusory or non-illusory, is based on the interaction of three essential factors; the local stimulation, the stimulus field forces, and the organismic field forces".

"If we now refer back to the six factors of unit formation as enunciated by Wertheimer we find that the first three factors' proximity', 'similarity' and 'common fate' are objective characteristics based on the forces in the field of perception. But the last two namely 'set' and 'past experience' are purely subjective or organismic, depending on the condition of the observer. Whereas the fourth the factor of 'goodness', if based on symmetry and balance will be objective and if based on the ease of formation or the pleasing feature of the formation will be subjective"

In 1942 Muenzinger suggested the term 'functional factors' to refer to those factors which derive primarily from needs, moods and past experience. In 1947 Bruner suggested the term 'behaviour determinants' to describe these factors.

Thus we can look upon the structural factors and the functional factors as two sets of factors, which are responsible for the perceptual organization. The gestalt phychologists were responsible to bring out clearly the importance of the structural factors. On the other hand the researches of Rorschach, Murray and Bartlett brought out very clearly the influence of functional factors in perception and other cognitive operations. We may give here some simple illustrations of the two sets of factors before we proceed to describe some of the experimental results. Just as a single black dot may stand out vividly in a group of white dots, similarly a single Negro may stand out prominently if he is in the midst of a group of white people. Similarly we find that a fair person may stand out prominently in a group of Indians. A fez cap or the Gandhi cap or the Mysore turban may stand out prominently when other people are not wearing any kind of head-dress. Here we find the operation of the structural factor leading to a perceptual organization. Similarly functional factors based on needs, will also influence the perceptual organization. A man who is rushing to catch a train or bus, or a student who is rushing to go to his class will not observe the restaurants on the wayside.

Some Experimental Results

In 1942 Murphy and others conducted a very interesting experiment which demonstrated how the intensity of a need lead to perceptual distortion. They showed ambiguous drawings behind a ground glass screen to two groups of College students. Students who were hungry perceived the ambiguous drawings more frequently as food objects than those who had just finished eating. They clearly demonstrated that this difference in perception is due not to the structural factors but to the differences in needs and motivations among the perceivers.

In 1947 Bruner and his co-workers obtained similar results with the size of coins. It was found that a group of slum children perceived the coins bigger in size than a group of children from business and professional classes. They asked ten-year old children to make size comparisons of various discs and coins. They found that the poor children tended to judge coins to be larger than the discs of the same size. They came to the conclusion that the social value of an object and the individual need for the socially valued object will influence perception.

"In experiments of a similar point R. L. Soloman found that children trained to receive a token which could later be exchanged for candy or other desirable objects would judge the token larger than it would have been judged before the training experience. The assessed over-estimation disappeared after the token was made no longer redeemable in terms of reward".

Krech and Crutchfield have enunciated two propositions which are of great significance:

Proposition (l) The perceptual and cognitive field in its natural state is organized and meaningful.

Preposition (2) Perception is functionally selective. It is a common experience that we tend to misinterpret the expression and manners of foreigners. We perceive their clothes, their language, their manners and customs as very peculiar, funny and even ridiculous. Often times we have found that people speaking the same language will look upon the regional differences in pronounciation as utterly ridiculous; probably the compliments are mutual. This is due to the fact that we tend to perceive sounds, movements etc., in an organized way and when they do not fit in with the organization with which we are familiar we tend to look upon them as peculiar and queer. In 1946 Asch conducted a very interesting experiment. He gave a list of traits and asked the students to write a description of the impression they had formed regarding the unknown person with those traits.

Thus we find that there is the influence of cognitive organization on perception. Sherif and other writers have made use of the concept Frame of Reference to explain such phenomenon. Buxton writes, "The frame of reference may be defined as the background of stimulation, which influences our behaviour in a particular situation. It may include external or internal stimuli other than the outstanding ones. It may include ideas or memories. But an important assumption is implicit in our simple definition, namely, that the effects of any given stimulus upon a person are not independent of the effects of other stimuli". Thus we find that the frame of reference brings out the significance of the factor that a system of functional relations influence our perception at a given time. We cannot explain perception merely in terms of stimuli, which impinge on the organism from outside. The internal factors also influence our perception. The important factor that we have to bear in mind is that there is organization, an integration involving both the external as well as internal factors. As Sherif writes "We shall then refer to the totality of external, and internal factors operating in an interdependent way at a given time as the frame of reference of the experience and behaviour in question".

A few illustrations from daily life will make this concept clear. When we say that a house is small we are obviously speaking not of this house as such; but of this house in relation to other houses. In other words we have a certain standard of reference regarding the size of the house. We are now comparing this particular house with the other houses. It is a familiar fact that a woman who is 5 ft. 4" tall is looked upon as a tall woman. On the other hand a man who is 5 ft. 4" tall will be perceived as a short man. Though the height in both the cases is the same we make this difference in judging their heights. This is because of the standard of height which is implicit and which integrates our present perceptions with our past experience. Because of the difference in the average height of the two sexes our perceptions of the height will vary. Similarly we perceive a man's behaviour as rude depending upon what we are accustomed as constituting polite behaviour and rude behaviour. What appears rude to a middle-class educated man will appear to be quite normal to an illiterate slum-dweller. Similarly an inappropriate remark may bring about embarassment in a formal social gathering, whereas the same remark may be highly enjoyable in another gathering. So it is not the stimulus itself that is significant but the occasion, the nature of the group, the time and such other factors will influence how we perceive that remark. It is a familiar fact that a student who gets 35% may feel very happy that he got a passing mark. On the other hand another student may feel humiliated because he has obtained only 58%. A poor man may feel that if he could only earn one hundred rupees per month he could send his children to 'the right school'. On the other hand a middle-class man may feel that even has income of four hundred rupees is not enough to send his child to 'the right school '. Thus the significance of the experience and the attitudes regarding success or the failure will depend upon the standard set up by the individual and the standard of the group to which an individual belongs.

"Distorted pictures take shape in the minds of men, but not because men are gullible by nature and not because they cannot see events with accuracy under proper conditions and orientations. Seeing things is not independent of the person's desires and biases or prejudices. In a complex social world where there are many alternatives to be noticed he is likely to notice those things which are relevant to his intentions and attitudes".

Pars Ram made a study of the Hindu-Muslim tensions in Aligarh in 1951. A few months earlier, in March 1950, there were widespread Hindu-Muslim riots at Aligarh and other parts of Uttar Pradesh. In an interview he asked both the Hindus and Muslims, "Have you heard or seen anything of the conflict between different groups or communities; *(a)* during the past year, *(b)* during the past six months, *(c)* during the past three months, *(d)* during the past week?" He found differences in approach between the Hindus and Muslims when they were answering this question. The Hindus reiterated that there were no incidents after March 1950 rioting. On the other hand Muslims gave instance after instance of mistreatment and discrimination by the Hindus. Thus it was found that there was little correspondence between the way in which the two groups perceived the same situation after the riots.

In order to illustrate the differences in perception of the same situation, which arise because of difference in our outlook reference may be made to the investigation by Zellig. In 1928 he got two groups of children to perform callisthenic exercises before their class-mates. One group was almost uniformly disliked by, their class-mates while the other group was very much liked by their class-mates. The experimenter had trained the 'disliked' group to perform the exercises in a perfect way. On the other hand the 'liked' group was trained deliberately to make mistakes. "When the two groups performed their exercises it was discovered that the audience reiterated that the 'disliked' group had made several mistakes. In other words the predisposition or readiness to dislike enables the person or a group to perceive errors. On the other hand a disposition to like has the opposite effect of over-looking the faults. It is a common experience that the step-mother always finds faults with the step-child. Similarly the mother-in-law always finds faults in the behaviour of the daughter-in-law. This has been very impressively brought out in Zellig's experiment. But it should not be inferred that prejudice always distorts our perception. Allport and his co-workers found at Harvard that persons with strong anti-semitic prejudices were more accurate in identifying Jews by their facial expressions than the persons with no such prejudice.

Murphy and Schafer conducted a very interesting experiment to show the effect of reward and punishment on perception. They used two ambiguous figures, which were presented momentarily. Each figure was so designed that a part of the picture could be seen as an outline of a human face. Every time one of these faces was perceived the subject was rewarded with money. Whereas the other figure when perceived lead to punishment by some of his money being taken away. By using this technique Murphy built up a strong association between certain visual patterns and rewards and between other patterns and punishments. In the test situation both the 'rewarded' and 'punished' patterns were combined into one picture so that either could be seen as the figure or as the ground. It is reported that 54 out of 67 perceptions were perceptions of the rewarded figures as faces.

It is a familiar fact that the mental set affects the perception. The writer recalls several occasions when he found well-settled Government officials remarking that the group of volunteers marching past the streets in 1942 as being composed of rabble and the waifs and the strays. On the other hand the same spectacle would make the patriotic and the nationalistic individuals to remark about the disciplined way in which the young men were moving in the procession. Similarly to the labour leader the contingent which is doing Satyagraha near the gates of a mill appears to be a disciplined, sacrificing and idealistic set of people whereas the same spectacle will cause the management representative to look upon them as ill-dressed, unkempt, disorderly persons who are preventing the disciplined men from doing their duty. When we look at the individuals who have been arrested on suspicion as being involved in some theft we are able to perceive pronounced marks of craftiness, irresponsibility, cruelty, and such other expressions in their faces. It is possible that many of them may be honest citizens who have been mistakenly arrested and who might probably be released later on. It is also a familiar fact that a person who is in a happy mood tends to over-look many things in a situation or in a person's behaviour whereas people who are in an angry or in an anxious mood tend to be very critical. In other words the perceptual structure for a person who is in a happy mood is simple and un-differentiated. Whereas the person in a critical mood will direct his attention to specific details in the perceptual field.

Anchorage

Thus we find that this concept of frame of reference helps us to understand many of the social phenomena. Sherif also used the term anchorage. An anchorage is a major reference point, which gives significance to the whole perception. The anchorage is the standard, which influences what we perceive. This anchorage may be due *(a)* to the structured nature of the external stimulus or *(b)* to the motives, attitudes, preoccupations and such other processes in the individual or *(c)* to the socially derived factors. The individual's attitudes or preoccupations or ambitions may become the anchorage, which re-organized the whole situation so that a particular pattern of perception may emerge. Similarly socially derived factors like group norms or group pressures may re-organize our perception. Sherif also draws attention to the disastrous influence on behaviour of the loss of stable anchorage. When we go to a strange place where we are unable to perceive any landmarks, where we do not have any kind of orientation, we become bewildered and we will become terribly frightened. Similarly lack of orientation may arise when our social ties with the beloved ones or with our groups are disrupted. A person may feel completely shaken up when his trusted friend betrays him or when some person near and dear passes away. Similar thing takes place in the economic situation when due to several reasons the prices of necessities shoot up the people become bewildered and they may find an anchorage in the revolutionary who calls them to attack and loot even the Government granaries. Under such Circumstances there may be a conflict between two anchorages.

The Influence of Age

Finally we may draw attention to some studies on the perception of facial expressions. As early as 1923 Gates reported a study regarding the growth of social perception using photographs of facial expressions. Photographs designed to express joy, anger, surprise, fear, scorn and pain were shown to children varying-in age from 3 to 14 years. 70% of the children at the kindergarten age were able to correctly name the picture showing laughter; but only less than half were able to recognize pain, anger and fear; none of the children recognized pictures depicting surprise and scorn. On the other hand of the seven-year old children more than 50% could identify anger. Fear was identified by more than 50% of ten-year old children and surprise by eleven year-old children. In daily life the children are able to identify expressions in familiar setting more correctly because of the situations, words etc.

Yet another aspect of social perception pertains to the way in which we perceive individuals as members of groups. The groups themselves try to develop unique ways of dressing etc. For example in India the people of different status have different types of dress. This leads to immediate perception of their status. Similarly employees in the army, railways, postal department etc., are given uniforms so that they could be immediately perceived as people with a certain status and function.

●●

9

Development of Mental Process

Some people, for example, equate high intelligence and the capacity to do school work. They often ask the question, why does Johnny who has a high I.Q. function poorly in school?" They assume that an intelligence test score is a complete predictor, encompassing motivational, emotional, and personality factors. A single score in a psychological study cannot tell us much about an individual; it is meaningful only in relationship to the total set of factors. Miss Iverson, a young teacher, is talking with the principal about her class. Results of the fall intelligence tests have just been given to the teachers, and she is attempting to relate her classroom experiences with the test scores. First, she is concerned about Fred. On the intelligence test he produced one of the highest scores in the class. However, she sees little evidence of high ability in his classroom performance or the completion of homework assignments. She asks, "Is it possible to have such an I.Q. and not really succeed in school work, and if so, why?" This immediately reveals a fundamental lack of understanding of what the score can tell us about an individual. She then looks at Susie's score, which is perhaps one of the lowest in the room, although still within the average range for Miss Iverson's class. Miss Iverson takes exception to this test score, saying, "Why, Susie is one of my best students. She always completes her homework assignments. She tries her hardest, and she gets good grades on most of the tests." How is Susie's comparatively low score on this test and her successful classroom performance explained? Can a child's I.Q. fluctuate? What would he revealed by a look at intelligence tests given to Susie in the past? Is it possible that Susie is in the category that is seldom mentioned with much concern, the overachiever? Has she, by dint of effort been able to establish a position within the class that is considerably beyond her indicated potential? Mr. Delavan, the principal, says, "Let's look at all the relationships that are involved. Yours is a sixth-grade class. We already have two other intelligence test scores on this group. Fred has always scored high in potential ability, and we have as yet to see any evidence of this capacity within the classroom. However, when we look at the home and family relationships and some of the physical factors related to his case, it becomes apparent that his I.Q. score is probably accurate and that we need to devise new methods of encouraging him to function.

As to Susie, her two earlier I.Q. scores were considerably higher and placed her in a more favourable position within the group. Since the I.Q. tends to fluctuate, Susie's highest score would remain most representative of her ability. No child is ever able to score higher than his capacity and perhaps Susie was not functioning at her best on the day of the test and hence produced a low score. A minimum estimate of a child's capability is the child's highest I.Q. score.

All teachers need help in learning to analyse the meaning of test scores more effectively. There has been far too much unsophisticated utilisation of both intelligence test and achievement test scores in the classroom. The scores can provide us with a measure of individual differences in performance in certain instances, but a single score should not be used to predict the potential of a given individual or to make permanent educational decisions. Perhaps a good policy for all teachers would be to recognise that test scores are most effectively utilised in the classroom in conjunction with other types of data. The teacher should also consider the child's personality, social development, physical makeup, concept of self as an achieving individual, and the way he has chosen to adjust to classroom expectations.

As we begin our study of intelligence, some questions might be posed. First, what do we mean when we talk about intelligence? Is there only one kind, or are there several kinds of intelligence? Is the I.Q. constant, or does it fluctuate? How much is the I.Q. influenced by heredity, how much by environment, and what can an individual do to improve the I.Q. scores of a given set of children? Are cultural factors present which prevent the child from scoring as well as he might? Does the "culture-free," "culture-fair" test really eliminate the factor of inequality of experience, and if we get a "culture-free" test, what kind of prediction does it give of functioning in schools that are closely tied to the culture? How is intelligence measured, and how often should it be measured in the schools? Is it possible for a child to raise his I.Q. through a change of environment, tutoring, intensive study, cramming, travel, and other types of experiences? How do emotions affect the intellect? Is it possible for emotional instability to affect the intelligence score?

CONCEPTS AND THEORIES OF INTELLIGENCE

All of us have some idea of what we mean when we say an individual is intelligent or lacks intelligence. Actual observations of behaviours are commonly used as indicators of intelligence. The speed and quality of an individual's response in a situation with intellectual components would certainly give us some idea of his intelligence.

The concept of intelligence has a variety of meanings, which are difficult to reduce to one simple definition. Intelligence has been defined as the ability to carry on abstract thinking, the ability to generalise, and the method of response to a problem situation.

Since Alfred Binet stimulated more psychological research studies on intelligence than any other single psychologist, perhaps we should begin with his definition of intelligence. Binet believed that intelligence was a combination of capacities that enabled the individual to adapt and be self-critical. In his general-factor theory, he included the abstract, mechanical, and social areas of behaviour as part of intelligence. Binet was concerned originally with establishing an accurate diagnosis of the intelligence of the retarded. His task was to determine whether certain students were ready for school experiences. In his tests, Binet stressed the examination of all types of capacities in order to determine criteria for stating that a child lacks the capacity to do scholastic work. In his original tests, in France, in 1905, Binet emphasised the primary factors of reasoning, comprehension and judgment. He proposed to set up a scale of tests to measure the development of the mental processes. Among the processes he included memory, images, recall, and attention. He established an age-scale approach to intelligence and eventually began to include children at all levels of intelligence. Binet made a most significant contribution to the study of intelligence. The construction of the intelligence scale and the emphasis on individual differences made a lasting contribution to the measurement of intellect.

The Binet test is based on an age-scale performance formula. The normal child passes all items at age six when he is six. The normal eight year' old has a mental age of eight, the normal ten year old a mental age of ten. Terman at Stanford University revised these scales in 1916, 1937, and 1960. The scale at the six-year level investigates vocabulary, differences in the meaning of words, ability to pick out missing elements in a picture, number concepts, opposite analogies, and maze tracing.

Lewis Terman first translated, revised, and published in 1916 a Stanford revision of the Binet-Simon test to be used in

English-speaking cultures. In this work Terman idefined intelligence as the ability to carryon abstract -: thinking and to utilise abstract symbols in the solution of problems. Terman's research seemed to indicate that vocabulary was the best single indicator of an individual's general intelligence. Terman's revision of the Binet was redone in 1937 and then again, posthumously, in 1960; it provided psychology with an extremely useful predictor of the capacity to utilise abstract symbols in a variety of cultural settings. Not all psychologists have accepted Binet's and Terman's definitions of intelligence. Some have developed intelligence scales based on different assumptions, such as the Wechsler and Thurstone assumptions. Charles E. Spearman began the transition from the general-factor theory to multiple-factor analysis, through his development of a two-factor theory of human capacity. He called the two factors involved in the intellectual performance of a task "G" and "S." "G" underlies all mental functioning and is considered the common denominator in all sorts of mental performance. This factor varies from individual to individual but is consistent in all aspects of a given individual's mental functioning. Since mental functioning in unrelated situations calls for different types of mental accomplishments, a specific factor must also operate in 'each situation. The factor US" varies not only from individual to individual, but also within the individual, *i.e.*, intraindividual variability.

L.L. Thurstone was an important figure in the early development of factor analysis. He believed that a single intelligence index was inadequate for the purpose of describing mental capacities. He believed that a great many factors entered into specific skills in a variety of combinations. Multiple-factor analysis was designed to discover and isolate fundamental human traits, and Thurstone was able to isolate the following factors: ability to verbalise, word fluency, number facility, memory, visualising or space thinking, perceptual speed, induction, and speed of judgment. The results of

Thurstone's Analysis of Intelligence appears in the Primary Abilities Test (PMA) published by Science Research Associates. The 1963 edition of the PMA covers grades K-1, 2-4, 4-6, 6-9, and 912. The purpose of the test is to investigate primary mental abilities considered most significant in schoolwork.

Numerous psychologists have suggested that there are really different kinds of intelligence. Thorndike said intelligence might be divided into three areas-abstract intelligence, or the ability to deal with the symbolic, *e.g.*, a word, a code, or a geometric figure; mechanical intelligence, or the ability to deal with concrete objects; and social intelligence, the use of psychological principles involving problems of human relationships.

A considerable amount of research has been based on factor analysis, and perhaps as many as fifty different intellectual factors may have been identified. J.P. Guilford has attempted through factor analysis to develop a unified theory of intelligence. He defines a factor as a unique ability needed to do well in a particular class of tasks. Guilford devised a theoretical model of one hundred twenty cells each of which represents a distinct ability. In his model, the factors are classified in three fundamental ways: processes, such as memory, cognition, divergent production, convergent production; contents or materials, such as symbolic, semantic and behavioural; and products, result of applying a particular operation to content, including units, relations, and systems.

Guilford speaks of several kinds of intelligence. Concrete intelligence enables one to use figural information: abstract intelligence affects abilities pertaining to both symbolic and semantic content; and, finally, social intelligence, which enables one to work effectively with people. Thus, Guilford presents us with a background which might eventually be used in' understanding the development and the measurement of intelligence.

From the phenomenological point of view, intelligence is a function of interaction. Intelligence is seen as dependent upon the richness and variety of perceptions possible for the individual at a given moment. Psychologists like Combs have pointed out that in most individuals the quality of perception hinges on the experiences the individual has accumulated. Combs believes that the quality of an individual's intelligence is affected by exposure to a wide variety of experiences that permit him to improve perception. Implicit in the perceptual approach is the belief that a person will not behave any more intelligently than he believes he can. A child who is convinced that he cannot do arithmetic and spelling or understand geography will act in accordance with his convictions. To enable him to utilise his capacities, his convictions must be changed. Intelligence then is considered as a product of the interaction of a person and his perceived psychological environment, and the intelligence test does not accurately measure the potential differentiations the individual can make, but his functional perceptions. The phenomenologist would, therefore, caution users to recognise the factors that limit perception, such as physical conditions, opportunity, experiences, exposure to symbolic events, and the individual's own goals and values.

MENTAL GROWTH

As a child develops, he makes increasingly complex adaptive responses to his physical and social environment. Mental development can be inferred from this behaviour. While children differ widely in both rate and pattern of mental development, developmental progression can be observed and, to an extent, measured.

Mental growth curves have been developed for infants. In one of the earlier studies of mental development, a group of thirty-one male and thirty female infants were studied. During the first two years of life yearly increases in point scores were noted, as shown in the accompanying graph. Scores indicate that the infants' mental development was rapid to about the ninth or tenth month, after which deceleration was evident.

Since tasks required on an infant test are of a motor and perceptual nature, behaviour growth in the early months of infant development has been demonstrated to have little predictive validity for the later tests of development of intelligence. To understand individual mental development, it is necessary to study longitudinally the development of intelligence in individuals. Curves derived from averages of the mental development of a group of children differ from repeated tests of individual children over a period of time. In any study of mental development, we should always be cautious about inferring the rate of development from mental test scores. These scores do not represent absolute measurements, the size of which are necessarily equivalent at each point on the scale. For example, the year's mental development, even in the case of an average child, may not be the same between any two successive years. The intellectual increment between the ages four and five is not necessarily the same size as the increments between eight and nine or twelve and thirteen. We do not have an absolute scale for measurement in the intellectual area.

Intelligence in young children is measured by functions considerably different from those used to measure older children. Thus, it is obvious that the scales are far from equivalent. On the Binet a four year old is asked to identify objects in a picture vocabulary, name objects from memory, discriminate between forms, draw opposite analogies, and do simple comprehension items. At age eleven he would be required to produce a design from memory', comprehend verbal absurdities, define abstract words, memorise a sentence, solve problems, and indicate how certain objects are similar.

The abilities being measured at eleven are somewhat different and cannot be assumed to be merely an extension or expansion of four year-old abilities.

Curves representing scores made on repeated administrations of tests may not reflect actual intellectual development. Courtis maintains that no single type of growth curve can adequately express the pattern of mental growth because of the nature of the items included on the typical intelligence test. These curves lack stability, he believes, because they do not measure a uniform function in a uniform manner. Courtis thinks adequate study of mental growth will come only when developmental progress in the performance of a single act is considered. Courtis indicates that we must get back to single-variable research if we are to study development accurately. Growth curves have value because they make it possible for us to note the child's rate of progress, spurts, plateaus, and regressions in relation to his unique pattern.

Researchers in mental development have attempted to determine the point of mental maturity when mental growth ceases. The finding of Terman and Merrill that mental age does not increase after the age of fifteen years has since been attributed to the limited ceiling on the 1937 Revision of the Stanford-Binet. Evidence based on a wider sample and retesting of the same population shows that the age of terminal growth may be twenty-five or even beyond. There are reports of gains in intelligence test scores at age twenty-five on the Wechsler-Bellvue, and even at age fifty on the Army Alpha and Concept Maturity Test.

Bayley studied five boys from the age of one month to twenty-five years. She found that each child had an individual pattern, and after the infancy period there was an underlying pattern of development constancy. These five boys were tested at twenty-five years of age and all had continued to improve in their Wechsler-Bellevue scores. Bayley indicates that the intellectual processes measured had not reached a ceiling, with fourteen out of fifteen participants continuing to show gains. Thus the issue of terminal growth in development of a cognitive process varies considerably from individual to individual.

Research in mental development has pointed to some general trends. During infancy and the preschool period, the abilities measured by intelligence tests are generally perceptual and sensory-motor in nature. Following this early stage of development, the lementary school years use tests in which abstract intelligence tends to be most important. Abstract abilities are highly correlated. During predolescence and later on, tests show a considerable differentiation of intellectual abilities. Thurstone domonstrated a differential growth rate for mental abilities. Perception, reasoning, and space abilities develop somewhat earlier than numerical and momery abilities, while verbal abilities develop more slowly.

In its time, The Harvard Growth Study was probably one of the most comprehensive longitudinal studies in the field of physical and mental growth. Some of the conclusions of Dearborn and Rothney are significant for our study of intellectual growth :

1. Physical and mental growth are essentially individual affairs. No two cases have been found to have exactly the same developmental history as indicated in terms of their deviation from the averages of groups of which they are members.

2. The relationship between physical measurements and mental measurements is so low that the knowledge of one does not enable us to predict the other.

3. The group mental tests used in this study yielded higher IQ's than the Stanford-Binet tests.

4. In general, children tend to remain throughout the period of their mental growth (to age sixteen) in the same classification as they were at age eight.

5. The use of different mental tests over the years indicates that each test is characterised by its own single and peculiar differences with respect to the problems of practise effect and its relation to individual test problems.

6. In general, individuals tend to Perform at the same level on verbal and non-verbal material which appears in group mental tests.

7. Mental growth, as measured by the type of group mental tests used in this study, continues much beyond the age of adolescence, although with markedly decreased rate after the age of twenty.

8. Complete substitution of non-verbal for verbal material in mental tests, would result in handicapping as many children as does the use of tests using verbal material only.

9. The possibility of using the percentage of growth based on the estimated maximum growth of the individual, or of an unselected group, in preference to the mental age technique, and the inadequacy of the commonly employed intelligence quotient, as an index of mental growth has been demonstrated.

10. The advantage of using individual growth curve constructed with the above described "growth unit" has been noted.

11. Performance on mental tests aoes not seem to be related in any way with pubescent growth spurt.

In a more recent analysis of the Harvard Growth Study, Ethel Cornell and Charles Armstrong found general patterns of mental growth occurred in spurts.

1. A single growth curve theoretically reaching maturity between the ages twenty-six and twenty-seven, maturity being indicated by a mean level of ability regarded as just above average.

2. Two growth curves-the earlier one terminating around age thirteen, the second one not becoming evident until age fifteen or sixteen, with maturity theoretically indicated at about twenty-three to twenty-four years of age, at a level slightly above average. (Female)

3. Two growth curves-most rapid development from nine to thirteen, the later period one of slow increase with maturity theoretically reached at age twenty-eight to twenty-nine years, at the highest mean level of ability of all the patterns.

Individual differences in mental growth are observed in early infancy. When compared to a group, the mental growth of children seems to remain at a fairly fixed position, unless some serious changes in conditions of development occur. The child who fails to grow consistently cannot be predicted. The growth curve of intelligence can be generalised, but without certainty because of individual differences in both rate and maximum.

Nancy Bayley pointed out some striking examples of variation in mental growth. One child in her study, Mark, varied from his highest score when tested at one month, to group norm at one year, to the lowest one-fourth at two years; he remained at this position relative to the group until age seven, then rose steadily until he received the following I.Q. scores: at nine years 117; at ten years 122; at twelve years 130. Gerald, the brightest six month old, was consistently slow from then on, scoring 85 at nine, 84 at ten and 76 at eleven. Charles rose from the twenty-fifth percentile at age one year to average intelligence at four years and scored 146, 149, and 153 at nine, ten, and eleven years of age. These scores would place him in the ninety-ninth percentile of the Stanford-Binet scale. In the Berkeley Guidance Study, one girl rose from the thirty-first percentile at twenty-one months to the ninety-ninth at six years. Another at the same age dropped from the fifty-fourth to the sixth percentile. These extreme variations are not the rule, but they do show the danger of trying to predict the IQ, particularly in the early years. The age at which an individual ceases to grow in intellectual ability may range from adolescence to the twenties. Increases in ability after the age of twenty appear to be horizontal-in other words, an extension of knowledge.

Our study of mental growth curves indicates the necessity for a revised conception of intelligence by the schools and teachers. It seems reasonable to suggest that the research presented should shake the belief of those who tend to think that a single intelligence test will predict a mental growth pattern.

To speak of the constancy of the I.Q. without considering the many factors which affect this score at different age levels is to disadvantage a great percentage of the children. The use, in the elementary school, of I.Q. scores for the development of homogeneous groups becomes a particularly naive concept. Constancy of the I.Q. can now be disputed, and teachers, administrators, and psychologists would do well to consider problems involved both in the measurement of intelligence and the mental growth curves before attempting any early classification of individuals.

FACTORS INFLUENCING INTELLIGENCE

Physical and Hereditary Factors

A natural endowment unfolds maturationally and sets limits on potential intellectual functions. This native intellectual endowment varies among individuals, and its maturation is hindered or assisted by the type of stimulation available in the environment during the early years. By comparing increments in mental and physical growth in curves of mental and physical development, Abernethy found that rates of physical and mental growth were essentially unrelated within the normal range of individual differences.

Olson and Hughes found a general correlation of all aspects of child growth when plotted according to derived age values. Shuttleworth, who worked with data from the Harvard Growth Study, also found that both boys and girls with early maximum-growth ages in standing height proved to be more intelligent than children with late maximum-growth ages. His findings support the evidence of Abernethy, Olson, and Hughes that growth proceeds in patterned fashion.

Cultural Factors

Investigations (conducted since the beginning of psychological research) tend to show that the higher the socioeconomic status, the higher the average intelligence of the children. The basic question is whether higher socioeconomic position helps to raise the level of intelligence or, more simply, whether intelligent people rise in the social scale? It has been demonstrated that the intellectual level of the underprivileged can be improved by raising their socioeconomic levels, especially if an early enough start is made. However, not every child gains and the degree of improvement varies greatly. The culturally rich home environment tends to lift the I.Q., some believe as much as twenty points, while the home environment devoid of positive cultural influences, may lower the I.Q. as much as twenty points. More adequate research on this issue is still needed. Some problems obviously exist in failure of the tests to equate for cultural differences. What does the upper socioeconomic group provide that results in a higher intelligence quotient? The most important factors appear to be nutrition, adequate medical attention, security (derived from freedom from economic deprivation), and exceptional educational advantages. A child, devoid of advantages in the area of food, clothing, physical and mental care, is frequently a culturally deprived child who scores low on the intelligence test.

During Second World War, armed services tested the intelligence of people from all sections of the country. Scores were found to parallel the amount of money per pupil spent on public schools in the various states. Men from states with high educational expenditures had the highest average I.Q. scores. Some argue that tests favour the upper socioeconomic group. This argument is made particularly by people interested in developing the culture-fair tests. The upper socioeconomic classes provide advantages, but psychologists believe that a "good psychological home" is more important than a home with just economic advantages, since every child has a unique set of nature and nurture experiences. The family atmosphere and family constellation for him are different from his brothers' and sisters'.

The nature and nurture, or, hereditary and environment problem is not easily solved. Most research studies on this problem are inconclusive. The best conclusion to date is that the influences of heredity and environment are inextricably interwoven, and, hence, it cannot be demonstrated that one is a more significant factor than the other.

Emotional Factors

Richards studied the development of a disturbed boy, with an emphasis upon the relationship of intelligence test performance at various ages to the child's life situation during the same period. He produced two sets of data, test scores and a history of the individual's adjustment at the time of each testing. The intelligence testing showed the following results on the Stanford-Binet at various ages: the average I.Q. for all testing WAS 124, the individual IQ's fluctuating between 117 and 140. There are four trends in the curve he obtained for IQ's between the ages of three and ten. First, a rise of eleven I.Q. points from age three to four, then a drop of thirteen points from five to six, a rise again of twenty-five points from six to eight years, and finally a drop of eighteen points from eight to ten years. The boy, referred to as Bobbie Jones, was studied at the Fels Research Institute.

Bobbie's father, a college graduate with an I.Q. of 127, was a businessman in moderately good circumstances. The child's mother had some college training and had done secretarial work before marriage. During the first five years Bobbie seemed to lead a normal life, perhaps with a greater than usual attachment to the mother, although he was also close to his father. The father played with and read to the boy frequently. During the period from five to ten, immediately preceding the rise in I.Q., the father, in addition to active business during the day, started a company, which took most of his evenings. Bobbie saw less of him, but they continued to have a good relationship. In the period from age six to eight, when the I.Q. rose twenty-five points, during the middle of the second grade, he was switched from a stern disciplinarian to a more sympathetic teacher. He continued with her through the remaining year and a half. Thus, he spent a year and a half of the two years during the ages six to eight with this teacher. At the same time, his father gave up his night work and started spending considerable time with Bobbie. Whereas, the father had formerly been tense and nervous because of the two jobs, he became a new man, more thoughtful and easier to live with. The Fels Parent Behaviour Scales applied to the home during this time showed that the home was always well-adjusted and democratic but that it now accentuated these characteristics. It was during this period that Bobbie's I.Q. was 140. Thereafter, in the later phase, the home atmosphere was more severe with less devotion, affection and support of Bobbie.

During the last phase, ages eight to ten, when the I.Q. dropped eighteen points, Bobbie's "understanding" teacher became a tired and inactive teacher and eventually experienced a tumor operation. Bobbie himself was promoted to the fourth grade, where he had a severe teacher. In summarising this history, Richard commented on the following points: the school situation, which fluctuated with the rise and fall in I.Q.; the role of the father, which fluctuated with the rise and fall in I.Q.; and the fact that the highest I.Q. was obtained during the period in which the child-centeredness of the home was at its highest.

Family Resemblances

A number of investigations on intelligence have explored the relationship between parents and their own children, parents and adopted children, twins, and siblings. Leahy noted that the coefficients correlation between intelligence in parents and adopted children correlated about .20 while coefficients for own children were generally found to be about .50. This might be considered a significant difference since adopted children are generally assigned by agencies on the basis of a fairly high relationship between the parents, the child, and the educational setting of the home.

Many studies have shown that in general we can demonstrate a correlation of about .50 between the intelligence of fathers or mothers and their children. This same approximate correlation has been found between siblings, who live in somewhat the same environment.

Correlations between the intelligence of identical twins who have developed from the same fertilised ovum usually range between .70 and .90. Newman, Freeman, and Holzinger found that the IQ's of twins who experienced the same environmental opportunities and education were on the whole closer than the IQ's of twins whose environment and education differed considerably. They also noted that identical twins who were reared apart were, as a group, more alike than siblings who were reared in the same home. Honzik suggested that there is a significant relationship, which changes with age between parental ability and the child's intelligence test scores. For example, she found a shift from a coefficient correlation of about .05 between mother and child at age two to a coefficient as high as .35 at age five and thereafter. Data from Skodak in the same figure show similar relationships between the adopted child's I.Q. and the true mother's education. Honzik concluded that the parent-child correlations reflect individual differences, which are largely determined genetically.

Shields did a study of forty-four monozygotic, or one-egg twins with exactly the same hereditary equipment. He found a correlation on intelligence of. 76 when these twins were reared together, a correlation of .77 when reared apart, and a correlation of .51 in dizygotic twins where hereditary equipment is not identical. This points to the strong influence of heredity.

School Factors

A basic question is the extent to which the child's intelligence is influenced by the type of schooling he receives and the time he receives it. There have been some investigations in connection with nursery school and the raising of a child's I.Q. Olson and Hughes found that nursery school children from privileged backgrounds did not differ significantly in intellectual growth from non-nursery school children who also came from privileged backgrounds Remember that these children came from homes where nursery school experiences could add little to the nurture of the child's intellectual development. In this study, then, no special intellectual effects were attributed to nursery school and kindergarten attendance for children who experienced adequate nurture in their own homes.

Skodak and Skeels did a study in an orphanage, which provided a minimum amount of stimulation and opportunity and in which, as a result, were many deficient and deprived children available for research. Two groups were used in this experiment—an experimental group and a control group. Both groups remained residents of the orphanage, but for several hours each day the experimental group attended a nursery school. Skodak and Skeels found the children who attended the nursery school over a period of twenty months showed on an average a gain of 4.6 points in I.Q., with some individuals showing larger gains. During this period, the control children showed an average loss of 4.6 points. It was also demonstrated that gains in social and emotional areas were much more impressive than I.Q. gains.

PREDICTING INTELLECTUAL GROWTH, CONSTANCY AND RELIABILITY

Tests are used not only to determine the present status of the individual but also to make predictions about his future growth and status. In order to make an effective educational decision, we need to have the best possible measures of intelligence available.

Anderson concluded that the prediction of later status from early test scores will be less accurate the younger the child is at the time of first testing or the longer the interval between tests. After the appearance of speech in the child, tests begin to have greater predictive value. In general, it could be concluded that indices for prediction of future development are quite ineffective until the age of six or seven. Some of this ineffectiveness is due to the fact that infant intelligence tests deal with functions that are quite different from those on tests used during the elementary school years and adolescence.

Intelligence is a developing function and the stability of measured intelligence increases with age. Bayley did a comprehensive study of the intellectual growth of forty children between ages one month and eighteen years of age, which showed wide differences in the growth patterns of individual children in the area of mental ability. The shifts were almost unpredictable, occurring at all age levels and in a wide range of abilities. In an article in the American Psychologist, in 1955, Bayley presented some of the different patterns of intellectual growth between one month and twenty-five years of age. The data clearly illustrate differential periods of acceleration in individuals.

Fostering the Growth of Intelligence

Factors have been identified as beneficial to the growth and development of intelligence. The psychological climate that the child is raised in is one of the most significant. The child needs to feel accepted and valued as an individual. His ideas; thoughts, and even idiosyncracies should be listened to, accepted, and discussed in a mature fashion. He should not be laughed at as immature because he has an idea that is different. Instead, a climate should be created in which having unique ideas and making contributions to family living are rewarded.

The implication is that the child must be encouraged to make choices, solve problems, and become independent in his thinking. This type of behaviour should always be reinforced and encouraged. It is also vital that the child be provided with materials and information which permit problem solving. He should be allowed to develop new uses for old tools, to see new relationships, to do creative writing, and to have access to a wide variety of experiences and hooks. The family needs to provide a system for the regular recognition of achievement in the intellectual area. There should be an atmosphere which provides stimulation for all kinds of broad intellectual growth and which permits the child to work on his developmental problems and developmental tasks. He should be made aware of the alternatives and then encouraged to solve the problem. Adults need to have faith that children can learn from the consequences of their decisions.

The growth of intelligence is dependent upon a stimulating atmosphere and a creative relationship with the significant adults both in the home and in the school.

Measurement of Intelligence

Schools and placement agencies have been interested in the subject of intelligence testing because of its implications for placement. Intelligence tests attempt to predict the individual's capacity to function generally in tasks that require scholastic aptitude. They measure the individual's ability to cope with situations requiring the exercise of mental processes and try to measure the ability to comprehend the situation, apply past experiences, and solve the problem presented.

Tests are designed on the premise that intelligence can not be measured directly but is inferred by the observation or evaluation of behaviour. Behaviour has been grouped at various chronological age levels and decisions have been made as to what is intelligent behaviour for a given age level. The measurements are converted into norms for representative groups of a population, so that eventually individuals can be classified as having intelligence like a typical five year old or eight year old. Some tests, like the Kuhlmann-Anderson, used a median score in an attempt to avoid the averaging of scores in determining the norms.

Most tests of intelligence place a premium on the ability to do abstract thinking through the use of symbols. Vocabulary also is of great importance in measuring intelligence. Many psychologists believe that as a child ' matures his vocabulary is the best single indicator of intelligence as related to scholastic aptitude. This measurement of vocabulary can be made in an individual test, such as the Stanford-Binet or the Wechsler Intelligence Scale for Children, or in group testing.

For a long time, test makers have been concerned with the problem of avoiding measuring only differences in cultural opportunities or education. An effective test item combines a criterion of novelty and universality. It is novel in the sense that it has not been encountered before by individuals taking the test; it is universal to the extent that all, at a certain level, may have had some experience related to it. The group intelligence test has frequently been shown to have a high relationship to the extent, level, and quality of the child's educational experience. When we speak of measuring intellectual potential or "native capacity," therefore, we need to be aware of the overlapping of hereditary and environmental factors.

The teacher and school administrator must be aware of some of the problems involved in the measurement of intelligence. We must not assume that tests with similar titles measure similar behaviours. We must always determine whether the group on which the norms for this test were based is comparable to the children we are testing. It is also pertinent to recognise that tests with dissimilar titles may actually measure the same thing. It is quite applicable in testing to ask, "What's in a name." Intelligence testing with children must hold to a definite concept about what the test items are seeking to predict.

Psychologists engaged in test construction also need to recognise the cyclic nature of mental development. At certain times mental development proceeds more rapidly than at other times. Those who use intelligence tests in the schools should recognise that children with identical IQ's seldom function identically. One reason for this could very well be that they had acquired their mental-age points on different kinds of abilities and tasks.

Inhelder and Piaget also found differences in the modal age of children at a given stage on two different tasks. The age level for attainment of a given stage appears to be specific to a task rather than general, so that the child can operate at stage two on one task and at stage one on another. The first individual intelligence test was the StanfordBinet. Originally devised for use in dealing with a specific problem in France, it was later revised by Terman and then Terman and Merrill, and has been used extensively in the United States. The test items in this scale have been arranged in the order of empirically determined difficulty from the age of two upward. Items are always scored on a pass or fail basis. There are six test items at each half-year level from age two to five and six items at each yearly age level thereafter until the age of fourteen. Thereafter, the test provides for average and superior adult levels. The items have been arranged so that the normal child of a given age is expected to perform at his chronological-age level. Retarded children perform below the level, gifted children above.

Another frequently used individual intelligence test is the Wechsler Intelligence Scale for children (WISC), which the psychologist uses diagnostically to determine areas of strength and weakness. The Wechsler test, developed and originally copyrighted by David Wechsler in 1949, abandoned the concept of mental age originally introduced by Binet in 1908. The Wechsler test has instead worked on the basis of a deviation intelligence quotient. The I.Q. is obtained by comparing the child's test performance with the scores earned by individuals in a single age group, not with the composite age group. This was done in an attempt to keep the standard deviation of IQ's identical from year to year, so that a child's obtained I.Q. does not vary unless his actual test performance as compared with his peers varies. The Wechsler test also permits development of a verbal, performance, and total or full-scale intelligence quotient. The verbal factors include information, comprehension, arithmetic, similarities, vocabulary, and digit span.

Group intelligence tests are frequently used in the schools to make educational decisions. Close analysis reveals that many mental-ability tests being widely used are really strictly measures of classroom achievement, while others are quite independent of classroom instruction. The person who selects or uses tests needs to know whit the test actually measures. Individual deviations due to motivation, fatigue, distraction and the like may not be noted in a group-testing situation.

The following list includes some of the tests now being used in the public schools:

1. **The California Test of Mental Maturity.** Authors, E. T. Sullivan, W. W. Clark, end E. W. Tiegs. Publisher, California Test Bureau. Grades four to twelve.

2. **Differential Aptitude Tests.** Authors, C. Bennett, H. Seashore, and A. Wesman. Publisher, Psychological Corporation. Items generally well. constructed and frequently used in grades eight to twelve to make educational and vocational decisions.

3. **Cattell Culture-fair Intelligence Test.** Author, R. B. Cattell. Publisher, Bobbs-Merrill Co., Inc. School learned skills not required. Grades three to twelve.

4. **Lorge-Thorndike Intelligence Test.** Authors, I. Lorge, R. L. Thorndike. Publisher, Houghton Mifflin Company. A promising instrument for predicting achievement in grade one. Series considered to be a well-constructed set of growth scales.

5. **Henmon-Nelson Tests of Mental Ability.** Authors, T. A. Lanken, M. J. Nelson. Publisher, Houghton Mifflin Company. Places a. great emphasis on reading; scores tend to correlate well with teacher grades. Grades three to twelve.

6. **Science Research Associates Tests of General Ability (TOGA).** Author, J. C. Flanagan. Publisher, Science Research Associates. Measures cultural understandings and reasoning with geometric forms.

7. **Otis Quick Scoring Mental Ability Tests.** Author, A. S. Otis. Publisher, Harcourt, Brace & World, Inc. Grades four to twelve. Places a great emphasis on school skills.

8. **School and College Ability Tests (SCAT).** Author, Educational Testing Service Staff. Publisher, Educational Testing Service. Glades four to twelve. A new test, which provides both a verbal and a quantitative score. Considered to be a good measure of school-learned abilities.

9. **S.R.A. Primary Ability Test.** Grades four to twelve. Authors, L. L. Thurstone and T. C. Thurstone. Publisher Science Research Associates. Grades four to twelve. Stresses reading, arithmetic arid reasoning abilities.

10. **Kuhlman-Anderson Intelligence Tests.** Authors, F. Kuhlman and R. G. Anderson. Publisher, Personnel Press. Grades three to twelve. Well-constructed and considered to be one of the better measures of growth available. Emphasises both reading and achievement skills.

Interpretation and Use of Test Results

Individual tests permit a more reliable measure of intelligence than a group test. The motivation and effort produced are more uniform when the child works under the supervision of an examiner in a one-to-one relationship, and the examiner can observe the individual to determine if there is sustained effort and attention.

You should always be aware of the standard error of measurement for any test you use. For example, an I.Q. of 90 should never be thought of as exactly 90, but rather between 80 and 100. The instruments used for measurement are not precise, and the scores should not be taken as exact measurements.

Test results should be considered in terms of the reliability of the instrument. Has it been demonstrated that the test yields approximately the same result when applied to a number of children of the type you are concerned about? Also, is the test valid in the sense that It has accurately predicted the kind of precise information that you need for the decision you have to make? Are the norms representative of the characteristics of the group that the child being tested has to compete with? Are the size of the norm population and the recency of the sampling adequate?

A reading disability frequently affects I.Q. test performance on paper and pencil tests. Some evidence indicates that where reading is accelerated, the I.Q. is equally accelerated, whereas the child who is slow or retarded in reading may produce an intelligence score lower than his potential. Such factors as those related to reading can be checked out through the use of the individual test. It is also possible to give nonverbal tests to clarify the effects of a reading disability. The culture of the individuat .his background, and previous set of experiences should always be considered to see whether they are comparable both to the norm group and to the groups with which he is currently competing.

In interpreting the results of tests, the culture and environmental opportunity of the child must be taken into consideration. The same score means one thing in a child from deprived circumstances and another in a child from a rich cultural environment. The teacher should remember that IQ's on group tests may measure abstract intelligence or scholastic aptitude and that this is only one type of ability. Group tests rarely measure the ability to work with things or with people, nor do they measure the ability I to solve many concrete, practical problems. The child who scores low on group intelligence tests generally has trouble with academic work. However, he may have a good ability in mechanical, social, artistic, or other" areas. Thus, tests may tell us not only what behaviour is present and what is not present but may give us clues for other types of investigation.

●●

10

Process of Individuality

Old issues in development, for years now dead and abandoned, involved such questions as "Which is more important, heredity or environment?" "Is walking a product of maturation or is it learned?" "Is personality inherited or is it merely an individual expression of the culture?" Such "either-or" questions now seldom occur to students of human development.

There are, nevertheless, strong disciplinary biases in the current literature dealing with personality and human development concerning the relative importance of biological and cultural factors. Progress, however, is being made on both fronts, and the mysteries of biological heredity are gradually being unlocked. As we saw previously, with modern equipment for technical research the sciences of genetics and biochemistry are making steady progress toward an understanding of the intracellular mechanisms of hereditary control and development. In light of present knowledge, on the other hand, no one would question the essential role of the environment in all aspects and phases of development.

The generally accepted viewpoint, then, is that all development, whether physical or psychological, is the result of the interaction between the organism and its environment. From the moment of conception there is implicit in the individual a developmental potential and a particular design in terms of which the vital interaction of organism and environment will take place. The nature of this potential and design is determined by the particular combination of hereditary determinants present. The genetic factor determines the nature of and potentiality for organic response to and interaction with the environment. Heredity lays down the overall pattern of development and the individual limits of developmental achievement beyond which the individual cannot pass.

Of equally vital importance is the quality and adequacy of the environment. Obviously, if the environment does not provide the essential stimulation-the necessary nourishment and nurture-for optimum response and interchange, the process of development cannot proceed to the level of greatest potential.

Mechanism of Hereditary Endowment

Individuality has its beginnings in the union of the ovum, or egg, and the spermatozoon, or sperm. Each of these original cells, microscopic in size, is made of a material called cytoplasm and a nucleus, or central body. The tiny nucleus of each cell contains 23 minute, active, threadlike structures called chromosomes. Careful study of chromosomes, under the most powerful magnification, has revealed that they differ considerably from each other, each being identifiable in terms of shape and size and each consisting of a great many gelatinous, beadlike structures called "genes" strung closely together. It is estimated that the chromosomes, with their constituent genes, packed within the fantastically small nucleus of the sperm of the father and the ovum of the mother, comprise all the hereditary material from which the individual generates. As the sperm enters the egg the two cells fuse and become a single fertilized ovum with a single nucleus containing 23 matched pairs of chromosomes, or 46 in all. Thus, with the union of the two germ cells two separate lines of heredity come together with a total of some 80,000120,000 genes, each with its determining potential. The process of the development of a new individual is underway.

The parent cells have a twofold function in reproduction. Together they initiate the most remarkable and dynamic event in nature: the assembly of a living body out of single molecules of proteins, carbohydrates, and other biochemicals. In addition, the parent cells control the specific design of the body. It is a design that will follow a pattern passed along a chain of inheritance, going back to the biological roots of this family. In a sense each new life actually has no definite beginning. Its existence is inherent in the existence of the parent cells and these, in turn, have arisen from the preceding parent cells. When any two parent cells unite they bring together a blend of the attributes of all ancestors before them. Thus, since all people are descended from a small number of early human beings they are linked by a common heritage.

Environment

The term "environment" is generally used rather loosely to refer to conditions under which human beings live and develop. Recently "the environment" has become a matter of common conversation and popular concern as people generally have been made aware of the growing problems of air, water, and land pollution. They have become generally more. concerned about the abandon with which the world's natural resources are being exploited and wasted and the shrinking areas of wilderness are being defaced and littered.

In our present discussion, however, "environment" is given a more limited and precise meaning. It refers more specifically to these conditions, both in the external world and within the organism Itself, which are sources of stimulation and motivation and which collectively are regarded as a determining or influencing "factor" in human development. Actually, in this connection the term is used in two rather distinct senses. First, there is the more general reference to the complex of conditions and influences common to a population, or to a particular segment of a population such .as a certain age group or a group presumed to be going through a certain phase of development. Environment in such instances refers to the totality of living conditions external to the organism and objectively regarded. Thus we speak of the home environment, the "impoverished" environment of a particular geographic area, the prenatal environment, the "enriched" environment of a certain nursery school.

We also use the term "environment" in a more individualized sense to refer to the particular constellation of conditions and stimuli, internal as well as external, that are actually affecting the development or influencing the behaviour of a particular individual at a given time. This is one's effective environment. The adequacy, in both the qualitative and the quantitative senses, of the child's food intake or the health care tendered him are important items in his effective environment. Attempts are often made to modify or enhance a child's learning environment by providing demonstrations or interesting materials for him to manipulate or challenging situations with which he must cope. The lack of adequate and appropriate stimulation of particular "institution" babies that have been studied, when matched with family-reared babies, appears to have been the chief determent to their developmental progress. They suffered "stimulus deprivation". Their individual effective environments were inadequate for optimum development. In general the socialization of a child is a matter of regulating and manipulating his effective environment to bring about desired change.

Interaction

The essential "facts of life" and development have been stated earlier in a number of connections. They, of course, are *(1)* that every individual begins life as a single cell containing a blueprint of developmental potentialities and a controlling and guiding mechanism essential for their realization and *(2)* that from the moment of that beginning in order for life and development to continue there must be a protecting, catalyzing, and nurturant environment. That the processes of development must somehow consist of the "interaction" between these two essential factors is generally assumed. But this interaction process—that is, development—cannot be observed directly, and it is generally not well understood.

But development begins immediately. The minute organism grows and becomes more complex. This means, of course, that the interaction, almost from the beginning, is actually between the organism as is and the environment (the organism at any point in time being a product of earlier interaction).

At a more macroscopic level there has been considerable theorizing concerning the nature of the process of interaction between organism and environment. From the strictly stimulus response (S→R) point of view, of course, the role of the organism in the process is hardly a matter of consideration. The response, and the consequent learning, or development, is regarded as the direct result of the stimulus. But as Hebb (1958) and others have insisted, the precise pattern of activity which follows stimulation is not a function of the stimulus alone, but also of what is going on in the constantly changing brain-the "mediating processes." In other words, the input, the stimulus, is modified in its effects by the mediating processes, and the mediating processes are modified by the stimulus. The resulting development, new acquisitions, takes place as it becomes related to or "subsumed" into the already organized but changing "structure." This is roughly what Piaget (1952) means by his term "assimilation."

Piaget has given much consideration to the processes of organism-environment Interaction. First, he distinguishes between what he calls "functions," the processes of interaction (development), which never change and "structures" (schemata, patterns, or organizations of behaviour), which continually change through the processes of development. He discusses two related processes of interaction (functions): assimilation and accommodation. Assimilation is the process by which "reality data are treated or modified in such a way as to become incorporated into the structure of the subject" (Plaget and Inhelder, 1969: p. 5). Whenever the organism uses something from the environment and incorporates it, the process of assimilation is in operation. The ingestion of food as it is changed by the organism and is incorporated into the organism is assimilation at the gross physical level. The organism and is thus changed. Likewise, stimulus input is modified by the mediating processes underway, and its meaning is assimilated into the cognitive structure, thereby bringing about a change (development).

In assimilation, then, the interaction consists of, or results in, a modification of the input and its incorporation into the structure—the physical body, or the "cognitive structure." Accommodation, on the other hand, comes about in a situation in which the input, the stimulating object, or problem situation resists action or manipulation by means of previously established patterns of manipulation, thus making necessary an adaptive change in approach. For example, the infant sitting in his highchair has previously learned, that is, has assimilated the fact, that his rubber ball bounces and rolls when he drops it to the floor. He has enjoyed seeing it bounce and roll many times. He now has before him not his ball but his dish of cereal. He drops it, first presumably with the expectation that it will behave like his ball, but it does something quite different. It "refuses" to fit in with his present cognitive structure. He is likely then to experiment with dishes of cereal-push them off from different places on his table, dropping them from different positions, throwing them with force; but they refuse to behave like his ball. He thus learns the differences in the properties of objects. His cognitive structure becomes enhanced through accommodation (adjustment) to the properties of a bowl of cereal. In Piaget's words:

> the child then tries to make all new objects enter into the schemata already acquired and this constant effort to assimilate leads him to discover the resistance of certain objects and the existence of certain properties irreducible to these schemata. It is then that accommodation assumes an interest in itself and that it becomes differentiated from assimilation, subsequently the two become more and more complementary.

As Piaget insists, the processes of interaction-assimilation and accommodation—are "invariant." The processes are the same through the various periods of the life cycle while the "structure," physical and psychological, as we have already seen, changes greatly, continuously, and in many ways. Development is change.

Infancy

The so-called crisis of birth consists of the sudden radical change in total environment to which the neonate must adapt. His living medium has suddenly become radically different. He is no longer enveloped and cradled in relative quiet and darkness in a fluid medium of constant temperature. William James described the probable immediate situation of the newborn infant as one of being caught up in a "big buzzing, blooming confusion" of air currents, pain, pressures, light, and noise. But what is important to his survival is the fact that he reacts in special ways to certain specific stimuli within that confusion, thereby initiating certain physiological functions essential to postnatal existence. He responds, for example, to a slap on the buttocks or to a chilling air current with an inhalation and a gasp, thus initiating independent breathing. In his interactions with this new environment there is both accommodation and assimilation. The human neonate is the most helpless and dependent of all creatures, but he is also possessed with the greatest potentiality and capacity for adjustment and developmental change. The long process of development toward the realization of that potentiality soon gets underway as he thus interacts with his environment.

Early Experience

During the past decade or two much research on the nature of infancy and infant development has been done with both animal and human infants. This has led to a much greater general appreciation of the importance of the infant's environment. Much emphasis is now being placed upon the effects of early experience and stimulation. For example, the work with primates at the University of Wisconsin has demonstrated that infant monkeys, under conditions of social isolation and with a deficiency of sensory stimulation show a lack of the tendency to explore or to examine objects in the immediate surroundings-behaviour which is characteristic of normal monkeys. These deprived animals showed marked ineptness later on as adults in social interaction, particularly in their sexual and parental behaviour (Harlow and Harlow, 1969; Suskett, 1965). In general, the evidence from animal research indicates that when an infant does not experience normal contacts and association with his kind it fails to acquire some of the patterns of interaction that are "universal" and characteristic of his kind. In other words, what has been regarded as instinctive and "natural" to a particular species, m certain instances, might better be regarded as "developmental tasks" which must be achieved by the individual at the appropriate times (critical periods) in interaction with others of the species (J. P. Scott, 1962, 1963).

Studies of human infant development generally indicate the crucial importance of early environmental stimulation, particularly with respect to emotional and social development. The early studies of "maternal deprivation" of infants reared in institutions showed that when compared with infants reared by their mothers in the normal family situation there was a marked developmental deficit in the institution children. Bowlby (1951), in his review of these studies, revealed clearly the dire consequences of the lack of interpersonal stimulation that the home-reared baby normally experiences.

Psychoanalytic theory has placed great emphasis upon the importance of the gratification the baby normally experiences as he nurses at his mother's breast. According to theory, emotional ties between mother and child thus become established. It has been further assumed that in this early experience IS formed the basis of the capacity to receive and to give love. The research evidence in general, however, indicates that affectional ties are not dependent upon the gratification at the breast of hunger needs. Other kinds of sensory experiences, particularly those from the tactile stimulation of close bodily contact and other tactile and kinesthetic stimulation incidental to normal infant care and handling, are believed to be important in this respect.

So intimately related to this basic "love" interaction between mother and baby as to be another aspect of the same unitary experience is what Erik Erikson (1968) calls a "sense of trust." He regards this experience, engendered as it is in the feeding relationship between mother and baby, as "the cornerstone of a vital personality."

> For the most fundamental prerequisite of mental vitality, I have already nominated a sense of trust, which is a pervasive attitude toward oneself and the world derived from the experiences of the first year of life. By "trust" I mean an essential trustfulness of others as well as a fundamental sense of one's own trustworthiness.

This "sense" arises from the fact that the neonate's "inborn and more or less coordinated ability to take in by mouth meets the mother's more or less coordinated ability and intention to feed him and to welcome him." The infant must learn to coordinate his getting with his mother's way of giving, "as she develops and coordinates her means of giving." To the extent to which this coordination takes place, along with the affective quality of the interaction, the baby's sense of trust is established.

The level and the quality of the mother's ability to fulfill her part in this vital interaction is dependent upon a number of factors, such as her personal development, her experiences during pregnancy and delivery, her feelings generally about babies, and her attitudes toward nursing and caring. The baby's congenital temperamental nature, as we have seen, can also be an important factor here.

At any rate, the establishment of this sense of trust is regarded as of far-reaching importance.

> But in thus getting what is given, and in learning to get somebody to do for him what he wishes to have done, the baby also develops the necessary groundwork "to get to be" the giver—that is, to identify with her and eventually to become a giving person.
>
> It must be said, however, that the amount of trust derived from earliest infantile experience does not seem to depend on absolute quantities of food or demonstrations of love, but rather on the quality of the maternal relationship. Mothers create a sense of trust in their children by that kind of administration which in quality combines sensitive care of the baby's individual needs and a firm sense of personal trustworthiness within the trusted framework of their community's life style.

Other writers also have emphasized the importance of the quality of infant care—of the mother being "warmly dependable" and understanding in the care and nurturance of her baby (I. D. Harris, 1959). When such quality is lacking in infantile experience a basic sense of distrust can be engendered. This may be expressed in individual personality as a tendency to withdraw from social contacts and to be insecure, suspicious, and generally unhappy.

Studies also show that general "mothering" and the physical and emotional nurturance thus received results in more marked social responsiveness than is demonstrated by babies who are deprived of such special mothering (Rheingold, 1956). It has been shown in Russian studies of infant care that the behaviours of the adult caretaker incidental to the process of alleviating the baby's hunger are the crucial factors in stimulating positive responses. Such behaviours as talking, smiling, and facial expressions, rather than the actual alleviation of hunger, stimulate smiling and positive emotional responses in the infant.

Congenital Equipment

Challenged by the current need to know the origin and source of group differences in cognitive ability, psychologists became interested more than ever before in what the human infant is like at birth. With what "standard equipment" does he come? And especially, what is the functional level of this congenital equipment and how does it change with time?

The traditional view, of course, was that during early infancy, the period of complete helplessness, the child was too immature even to learn. His functional development was thought to be entirely dependent upon biological maturation. And even though he was equipped at birth with sense organs, they were completely nonfunctional at first. The tiny organism must first become "ready" through the process of biological maturation before he can learn to use his senses.

Infant Capabilities

Social change—changes in vital problem areas of national and overall social concern in recent years—has had a marked influence upon trends and areas of research in developmental psychology. Demands for social and racial "equality," for example, have stimulated much reexamination of longstanding theories about the nature of mental ability and the factors, which underlie differences in children of different social and racial groups in their readiness to meet and cope with school demands and expectations. Such group differences have long been observed. The unresolved question as to the origin and source of these differences has been a factor in the shift in research interest to the study of infancy. There is general agreement among researchers in infant development that young infants possess considerably more competence than was formerly believed.

Perceptual Development

The fact that learning, the effect of exercise, is a continuous aspect of development from the very beginning has already been emphasized. Infancy is a time of rapid learning. But recent and ongoing infant research has shown that infants are much more competent in the use of their sensory equipment than was formerly thought. Researchers are now interpreting their results to mean that perception does not develop through a "process of construction," that is, that the infant does not gradually build an ordered perceptual world through repeated and varied sensory experience, but rather that "infants can in fact register most of the information an adult can register" immediately (Bower, 1966). Visual discrimination interpreted as perceptual ability was found in very young infants in the fact that they exhibited a "preference" for (fixated their eyes for longer periods of time upon) certain visual stimulus patterns (Fantz, 1958, 1963). It was further found that an infant "at any age" can discriminate visually between solid and non solid objects, and that by the second week of life "an infant expects a seen object to have tactile consequences" as it is made to approach his face (Bower, 1971). From the results of many ingenious experiments Bower reached the conclusion that "in man there is a primitive unity of the senses with visual variables specifying tactile consequences, and that this primitive unity is built into the structure of the human nervous system". This is to say, of course, that the awareness of the properties and attributes of "solid" objects, including their "tactile consequences," does not have to be acquired (learned) through experience, but is a congenital ability—that such awareness is immediate and unlearned.

This, of course, is contrary to much of the earlier accumulated evidence concerning this basic cognitive process which we call perception. Ausubel (1958) quite clearly states the more widely accepted point of view:

> Since all perceptual and cognitive phenomena deal by definition with the contents of processes of awareness, they cannot always be inferred from overt behaviour. Behaviour, for example, frequently reflects the organism's capacity for experiencing differentially the differential properties of stimuli. Nevertheless, since all differential psychological experience preceding or 'accompanying behaviour does not necessarily involve a content or process of awareness, we cannot always consider it perceptual or cognitive in nature. Several examples may help to elucidate this distinction.
>
> We have previously shown that a young infant will follow a patch of color across a multicolored background. Will cease crying when he hears his mother's footsteps, and will respond differentially to various verbal commands. Conditioning experiments during infancy also show that the child is able to

> "differentiate" between different sizes, colors and shapes of objects and between pitches of sound. Does this constitute evidence of genuine perception...?

...it is reasonable to suppose that much of the sensory experience impinging on the infant is too diffuse, disorganized and uninterpretable to constitute the raw material of perception and cognition despite evidence of differential response to stimulation. Clear and meaningful contents of awareness presupposes some minimal interpretation of incoming sensory data in the light of an existing-ideational framework. In the first few months of life not only is the experiential basis for this framework lacking, but the necessary neuroanatomic and neurophysiologic substrate for cortical functioning is also absent.

It is quite clear from the foregoing that this difference in viewpoint is very largely one of semantics and interpretation. The currently active researchers in the field of infant behaviour and capabilities have contributed and are contributing much to the general understanding of early human development. Because of their findings the whole concept of readiness to learn, for example, has undergone drastic revision. However, the question of whether young infants have the unlearned ability to perceive depends upon what one means by the term "perception." In the tradition and universally accepted definition of the term, perception is a cognitive (a knowing) process involving the conscious awareness of the properties and attributes of an object or a situation, which can only have been gamed from previous experience. Such an act, or process, of "knowing" is not congenital.

If, on the other hand, the term "perception" is construed to mean simply sensory discrimination, then in that sense infants may be said to perceive almost from the time of birth. Even Bower's finding interesting as it is, that a baby will make protective responses to the rapid approach toward his face of a "solid object" does not necessarily involve a conscious process of awareness (an actual apprehension of danger) of the "tactile consequences" of the approaching object.

In any event, the neonate does react to sensory stimulation, and the fact that he does something means that whatever he does brings about developmental change. The structural unit involved in what he does, simple as it may be, is exercised. Change (learning) takes place. As Piaget observed, during the first weeks of life the baby exercises his various congenital sensori-motor schemata. Their functioning becomes stronger and more precise. Through exercise his seeing mechanism functions with ever-increasing clearness, and things touched and grasped become things seen. He thus begins to experience the properties of objects. His experience of his noisy, brightly colored rattle in his mouth is different from his experience of his fist or his thumb in his mouth. Through such experience he is "coming to know" about things and their properties as different from himself. The basis of his ability to discriminate cognitively between what is a part of himself and what is not himself, the beginning of self-awareness, presumably is thus gained early in his experience. Thus development is an integrated process. The baby's cognitive development (coming to know) is not something separate and apart. It is dependent upon, and indeed is an aspect of, the development of the reaching-prehensile-manipulation pattern. The affective experiences resulting from failure and frustration, or from success in grasping and handling result in changes in attitudes and emotional patterns. Thus the infant develops cognitively, behaviourally, and emotionally from his simple sensory experiences and his own overt responses to them.

Development in all its aspects proceeds at a rapid pace during infancy. Even as early as 2½ months the baby is especially responsive to the human face. He shows an increasing need for association, to be with others. By age 4 or

5 months his "affiliative" need is obvious. He distinguishes between familiar and unfamiliar people. He definitely recognizes his mother and is likely to demand more and more of her attention.

Achievement

During the earlier period of his complete dependency, as we have seen, the baby has enjoyed what has been referred to as "volitional independence" (Ausubel, 1958). What he "demanded" he usually got. He displayed what looks like a "sense of omnipotence." But as he gains more and more ability to interact with his environment and to do for himself there is a concomitant change in the character of his relations with those about him. There is a growing degree of interaction, behavioural and affiliative, and a lessening of demand on his part, and subservience on the part of others.

By the middle of his first year the baby has gained considerable command over his body. He can roll over from back to stomach and can sit up momentarily. By this time also he is likely to have learned to crawl.

Recent work on the perceptual and learning capabilities of infants suggest that by the end of the first six months an interesting aspect of visual perception is clearly evident. In one experiment it was found that 6month-old babies could not be prevailed upon to crawl across an area covered with heavy glass arranged so as to give the illusion of a drop-off to a dangerous depth. The infants apparently perceived the drop-off or "cliff" effect, which meant danger to them (Gibson and Walk, 1960).

The second half of the baby's first year is a period of signal behavioural achievements. There is a great deal of handling and manipulation of objects within his reach. He seems now to grasp things, not just for the sake of grasping. His actions are more "centered on a result produced in the environment . . . the means are beginning to be differentiated from the ends". He begins now to show some anticipation of the consequences of his acts—the beginning of "intentionality." He can also use his hands more expertly, both at the same time.

By his eighth month the baby is able to respond to more than one person at the same time. He likes to be "bounced" and handled, but may easily become overexcited. He vocalizes happily to himself, imitates sounds, and is beginning to respond to his name and to some other words such as "no."

By the time he is a year old, simple patterns of manipulation have become coordinated into more complex acts performed with the definite aim of attaining a specific end. He thus puts to work by intention a series of single schemata now coordinated for a new purpose. The "accommodation" aspect of his adaptations is especially apparent as he definitely begins to differentiate self from not self and is able to search, to a limited degree, for a "vanished object." He also shows evidence of an implicit conception of causality. He appears to foresee events that are independent of his own acts.

In his social responsiveness the baby can now play "peek-a-boo" and "patty cake." He may be saying "ma ma" and "da da" with definite reference to his parents. He notices differences in his mother's tone of voice when she approves or disapproves of his behaviour.

By this time also he may be beginning to toddle about. He is now in a transition phase of development. He is moving on from his babyhood.

Infantile Individuality

During this transition phase the integratedness of development and the interdependence of some of its signal aspects become especially evident. For example, both locomotion and speech learning are major developmental tasks of this time. Both are especially important to the child's general cognitive and social development. In many cases it appears that two tasks of such prodigious proportions cannot, in the child's economy, be given full attention at the same time. One of them for a period gains preeminence and spurts ahead while the other lags behind. The child may seem to be concentrating on learning to walk while his speech learning is neglected.

We should at this point be reminded again that every infant even at birth is a person in his own right. Regardless of whether our focus is upon the relatively profound dependency and helplessness of the human infant or whether it is upon his remarkable level of capability, which is only currently being fully recognized, we realize that babies are not all alike. Each is by genetic endowment unique, and with his individual pattern of congenital potentialities and temperamental inclinations he is born into a family situation, which is like no other. His original individuality, moreover, is likely to become enhanced as it interacts with his effective environment. We have noted how the child's realization of his potential for cognitive development is dependent largely upon the quality and adequacy of the stimulation he receives, and how his achievement of a basic sense of trust, together with a basic experience of a reciprocal love relationship, for example, is dependent upon the manner and the readiness with which his need and ability to take nourishment ,at his mother's breast is met by his mother's ability and dependability in giving him that nourishment and satisfaction. Human individuality begins at the very beginning.

Shift in Bodily Locus of Gratification

During the dependency of early infancy, as we have seen, the baby's psychological life presumably consists largely of the experiences and satisfactions associated with food taking, the cutaneous and kinesthetic experiences of being handled and cuddled, the relief from hunger, and the primary pleasure of sucking. This is the "oral stage" described by Freud, a time when life is centered in oral satisfaction and related experiences.

With the achievement of locomotor freedom and his expanded perceptual-motor capacity the child's sources of gratification also become broadened. By this time the child ordinarily has been weaned from the breast or bottle with the consequent loss of much of the oral satisfaction he formerly experienced. At the same time his mother's increasing concern for cleanliness in relation to his eliminative functions, with her bathing, diapering, and other "caring" procedures connected with them, tends to center the child's attention upon those activities and the sensory pleasures derived from such stimulation of the anal region. The eliminative processes them selves and the relief they give are primary sources of pleasure. These various factors in the child's experience conspire during this period of toddlerhood to give the anal region precedence over the mouth as the main bodily locus of gratification.

Toddlerhood

With the achievement of independent locomotion the child's effective environment becomes greatly expanded. As he moves about he is confronted moment by moment with a new array of objects, which to him are exciting and attractive—things to be reached for, grasped, and manipulated. Without experience he has no basis for judgment as to what may be grasped, mouthed and manipulated, thrown down, and what may not be touched. This he must learn. It is one of the earliest learning problems in his socialization. He learns these and many other lessons of conduct through interaction with the human aspect of his environment, particularly his parents.

The effective environment of the toddler, thus, has rather precipitously become much more complex. It is two-sided. There is the world of things, which holds much promise of the gratification of an intrinsic tendency within his to seek stimulation, to "take in" his world. The other aspect of his environment, the human side, has suddenly become repressive, limiting, frustrating. In his interactions with this two-sided environment, the child learns something of the nature of reality and his own relation to it. His ego develops.

Drive To Explore

Perhaps what most characterizes the toddler child is his drive to explore his environment. His concentration in this reaching-grasping and manipulating activity is most intense. His approach to objects within his reach is one of "irresponsible impulsiveness." In this need to explore and manipulate he demands freedom to do just that-freedom to exercise his developing "executive independence," the ability to do 'for himself. With seeming abandon he moves from place to place grasping and manipulating objects, sensing their properties. He now sees things from new perspectives. He gains new sensory experiences of them. Some things yield readily to his grasp, allowing him to smell and taste them and more fully to "come to know" (assimilate) them. Other things resist his pull upon them. He must adapt his behaviour differently toward these objects (accommodation). Thus in such varied experiences an important aspect of perceptual development, which Piaget calls "the elaboration of the object," is rapidly taking place.

The child's demand for free access to his surroundings, however, often leads to conflict. At his level of maturity the child is not ready for such free and complete access. In our Western culture his surroundings must somehow be protected from his impulsivity. Restraint of some kind is usually resorted to. His freedom is curtailed and his "socialization" gets underway.

Emotional and Social Development

As those about him become restrictive, frustrating, even pain-inflicting, the child learns new lessons about relations and interactions with others. He must also revise his sense of self. In a sense he is in a crisis situation. His former "sense of omnipotence" is being frustrated. Some children react with the commonly observed "negativism" of toddlerhood. He too says "no no," and anger tantrums are not uncommon.

During this conflictful period of transition from babyhood, the child is also likely to fluctuate from time to time between a strong bid for independence and a clinging dependency. At times he may need the security of being "loved" and cuddled like a baby. It is important that his earlier acquired sense of trust in his environment not be impaired here. Dependability and consistency of treatment, with love and with respect for his individuality on the part of parents, is important during this period of socialization.

Toilet Training

The eliminative functions were discussed in some detail. The gaining and exercise of control of elimination is generally one of the most difficult developmental tasks of the early preschool period. Children differ widely in the ease of its achievement and the time required. This is frequently a problem of great concern for parents. The toddler in his preoccupation with things outside himself is developmentally unready to show any concern or to assume the least responsibility in relation to the time or place for the exercise of his eliminative functions. For him these functions operate automatically, as they have done from the beginning, under the control of the autonomic division of the nervous system. The child is completely oblivious of the concern of his parents for the maintenance of standards of cleanliness and order in the home. And, of course, parents usually take this for granted.

But soon the problem of elimination training becomes an uppermost concern of the mother. There is no set time when such training should begin. Knowing when the child is ready physically and psychologically is most difficult. To begin training procedures too early is likely only to prolong the process. One can only be alert to the signs of readiness, which are not easy to recognize. Physical readiness is a matter of being able to exercise sufficient muscular control to stop natural release. When the child is able to communicate by signal—by word or sound or look-that there is need to eliminate, and when he begins to show his ability to understand and accept the suggestions of adults, he may then be regarded as psychologically ready to learn voluntarily to control and to regulate these functions.

The very erratic course of progress made by our subject, Sally, in gaining night-time control. This child was reared according to the "self-regulation" regime. The regulation of her eliminative processes as well as her feeding schedule were largely self-initiated rather than imposed upon her from without. According to the record, no "training" of any sort had been initiated at age 18 months. At age 27 months, although she still wet her bed nightly, Sally had gained good daytime control. Although she was "taken" periodically and was encouraged to indicate her need, no pressure was ever used and the only reward she received for her successful self-regulation were the words "good girl." Sally's record illustrates dramatically the wide individual variation that must be expected in the achievement of voluntary sphincter control; ultimately, when success is achieved, it may occur rapidly.

Feeding

As we earlier noted, the feeding of the young preschooler is often a "problem area" for parents. It is an activity in which the child shows strongly his need to be independent. Because of the usual "mess" the child makes in his efforts to do for himself, and also because of the parent's need to insure the proper amount of food intake, the parent too often "takes over." Thus the child is denied the fun of self-feeding. He is likely to lose interest in the whole feeding procedure.

Feeding problems are usually created. Feeding time is another situation in which the interaction between the child and his parents can range from a happy and positive experience to one fraught with emotional conflict and frustration for both.

The Preschool Child

There is no definite age or point of development separating toddlerhood from the following three years, which we call the "preschool period." Nevertheless, there are certain distinctive developments, certain changes in the prevailing interests and activities and in the nature and quality of child-parent relations, and related problems through which the child works, which particularly characterize the later years before he enters school. It is a time of great importance in the child's social and emotional development.

Motor Development

In terms of overt motor behaviour there, is generally a vast difference between the 3-year-old and the 6-year-old. At the end of his first year of upright locomotion the child's walking is still not a "thing of ease and vigor." However, he is on the verge of a period of great achievement in skills involving both pairs of limbs. Soon he apparently comes no longer to need to consciously direct and control his walking as such. He can forget that he is doing the walking. He now walks to get someplace or to do something. His eye-hand coordination becomes more precise, and a great many skills, both manual and locomotor, are developed.

Physical Growth

The preschool period should not be regarded as a "stage." It is, rather, a span of time, covering roughly the third, fourth, and fifth years of life, during which rapid developmental progress continues to take place. Physically, the rate of growth—increase in overall body size, as measured by units of size gained per unit of time, although rapid, is gradually decelerating. In terms of actual measurements, at the end of his second year the child is most likely to measure 37 or 38 inches in height and to weigh about 33 pounds. Three years later the averages are around 46 inches in height and 45 to 46 pounds in weight.

Cognitive Development

Inseparable from the achievement of the skills for the control of the body and its parts in relation to the 'things and the conditions of the environment quite obviously is the acquisition of knowledge (cognition) about those things and conditions. Especially during the periods of development with which we are concerned in this chapter, the development of motor skills and cognitive development are, in' a real sense, "one of a piece." A motor skill is the ability to manage effectively the body and its parts in relation to the environment. This ability involves the awareness of the body and its limitations, as well as a knowledge of the nature of the environment the objects and conditions with which the skill is concerned. Knowing an object comes through its sensing and manipulation, and as that knowledge grows, more and more proficiency in bodily management develops for its further examination and manipulation. Thus the child's coming to know about the world about him (cognitive development) and the development of the motor abilities and skills in dealing with it are actually one and the same process. As we study separately one by one the various facets of development we tend to lose sight of the unity, the integrated nature of human development. Even at age 3 the child is usually well along in what Piaget calls the "period of concrete operations." He is rapidly learning words and their use in speech. He can manipulate word symbols that represent the environment about him arid thus communicate with others. His cognitive processes are still "preoperational," however. He is incapable of conceptual thought. During the latter portion of his preschool years the child's thinking is intuitive in nature in the sense that it is still lacking in logical analysis. It is based upon immediate, unanalyzed impressions of the objective situation. It is dominated by subjective perceptual judgments. For example, when confronted with two like glasses filled to the same level with water he will correctly observe that each contains the same amount of water. But after he has observed one of the glasses being emptied into a taller, slenderer glass, he will now say that the tall, slender glass contains more water than the other. His judgment is not a logical one based upon observed relations and related events. It is a prelogical schematization of perceptual data, an incomplete intellectual construction. The child at this level is not aware of his own mental processes. He does not think about his own thoughts. He "acts only with a view toward achieving the goal; he does not ask himself why he succeeds". Development through this intuitive period, however, leads to the threshold of operational functioning.

Social and Emotional Adjustments

The preschooler becomes more and more observant of the activities of those about him. He notes differences and makes comparisons between individuals of different ages and between the sexes. He tries to comprehend the various roles people play in life and in imagination "anticipates" his own roles and areas of future functioning. The little boy becomes clearly aware that he is a boy and will become a man, and the girl is equally conscious of her role and her destiny as a female.

The child at this age is likely to become more or less preoccupied with that area of behaviour. He may become especially curious and imaginative. He may also experience special genital excitability. But as Erikson points out:

> This "genitality" is, of course, rudimentary, a mere promise of things to come; often it is not particularly noticeable. If not specifically provoked into precocious manifestation by especially seductive practices or by pointed prohibitions and threats of "cutting it off" or special customs such as sex play in groups of children, it is apt to lead to no more than a series of peculiarly fascinating experiences which soon become frightening and pointless enough to be repressed.

By the time the child is 3 years old, however, in addition to what he has experienced in the process of socialization, much human behaviour, personal, interpersonal, and emotional, has transpired before his eyes and in his hearing. He has "incidentally" observed a variety of emotional interaction between his parents and other family members. And the things he experiences and observes in his home are what family living and its relationships mean and will continue to mean to him. Normally the child has become strongly "attached" to both his parents. By this time the little boy is beginning to identify with his father, the little girl with her mother. Whatever the child has observed in the behaviour of his model tends to have positive meaning for him and he tends to emulate it. He wants to "be like Daddy."

But his love for his mother is said, to begin to undergo a change in quality during this period. In terms of psychoanalytic theory, as was noted particularly, the child at this time is entering the phallic stage in his psychosexual development. He is becoming enmeshed in the so-called "Oedipal situation" in which he is said to become a rival of his father for the erotic love of his mother. His father, the only obstacle to his complete possession of his mother, is very large and powerful, hence the child wishes his father's death. Such unworthy wishes bring feelings of guilt and fear. "Castration anxiety" may now beset him-a fear of retaliation by his powerful rival. The resolution of this intolerable situation comes about through repression and a strong identification with his father-becoming like father means gaining "fatherly privileges."

Individuality among Preschool Children

Earlier in this chapter the obvious fact was emphasized that even though human beings possess many traits and characteristics in common, children at birth vary widely in the strength and intrusiveness with which these common features manifest themselves. Hence in his total patterning and combination, each infant at birth is unique. He is an individual, a person in his own right, different from every other. There was also the reminder that each and every infant, with his congenital uniqueness and with a great potential for development, enters a parental situation, an overall effective environment, which is also unique, and that through the interaction of these two uniquely patterned factors development in all its aspects immediately gets underway.

What happens as the various behavioural features and predispositions with which the baby is born develop depends in each instance upon such factors as its relative strength, kinds and degrees of environmental nurturance or neglect, and pressure or discouragement it receives, particularly during the period of infancy and early childhood. The child continues to be a unique person, but the eventual patterning of his individuality appears to be a matter of development. A child with a congenital predisposition to be highly active physically and with good potentiality for mental development, for example, could be ready to enter school at age 5 or 6 as a very capable beginner, with self-confidence, spontaneity, and full readiness to meet the expectations and cope with the demands of the school situation. However, with a different kind of care and handling, perhaps with many restrictions upon his freedom, vigor, and spontaneity of expression, this same child could have been entering school with limited ability to act on his own initiative or to be readily disposed to hostility and aggression when frustrated or when things do not go according to his expectations and therefore not well fitted for the new experiences of school.

The determining factor in many aspects of personal, social, and emotional development during this early period, then, appears to be the general, predominant quality of the emotional interaction between parent and child. Whatever the direction of development any of the child's congenital predispositions may take, the evidence suggests that by the end of the pre-school period his basic personality structure—his individuality—is pretty well modeled and established.

Role of Conscience

Presumably the preschool child would already have developed conscience (Erikson 1968, p. 119). The socialization process has been, to varying degrees, effective. The child would probably keenly realize that the feelings and the wishes he is experiencing during the Oedipal crisis are contrary to his parents' teachings about "right and wrong," what is "evil" and what is "good," and he may become frightened by his own thoughts and feelings. Erikson (1968) regards conscience as the "great governor of initiative" ("initiative" is his descriptive term for the phallic phase):

> The child, we said, not only feels afraid of being found out, but he also fears the "inner voice" of self-observation, self-guidance, and self-punishment, which divides him radically within himself; a new and powerful estrangement. This is the ontogenetic corner stone of morality. For the conscience of the child can be primitive, cruel, and uncompromising, as may he observed in instances where children learn to constrict themselves to the point of overall inhibition; where they develop an obedience more literal than the one the parent wishes to exact; or where they develop deep regressions and lasting resentments because the parents 'themselves do not seem to live up to the conscience which they have fostered in the child. One of the

deepest conflicts in life is caused by hate for a parent who served initially as the model and the executor of the conscience, but who was later found trying to "get away with" the very transgressions which the child no longer tolerates in himself.

Physical Development

The only suggestion of latency during these years is, of course, in the physical growth of the reproduction system. As was noted in Chapter 4, there is much diversity and change in the nature and rate of growth in the various tissues and organ systems of the body. In overall body growth, however, this period, roughly ages 5-12 years, is one of steady progress with relatively little change in rate and with no significant difference between boys and girls in either height or weight. Height as a rough average, with much individual variation, at 5 years is about 43 inches. At age 12 the average is in the neighborhood of 59 inches-an average gain of roughly 6 inches. In weight the average at age 5 is near 42 pounds. At age 12 the average is about 87 pounds, more than double the weight at the beginning of the period. At age 12 the girls are beginning to take the lead in physical growth.

Period of Elementary School Attendance

The descriptive term "latency" is often used to designate the period of elementary school attendance, a span of about eight years. In most areas of development, however, a great deal of change takes place. As is true of all the segments into which we arbitrarily divide the years from birth to adulthood, this is a time of marked and steady development. Generally speaking, it is in no sense a stage of latent or arrested development.

Motor Skills of the School Years

The elementary school years are the physically active years. The child is interested in the use of his body and in the exercise of the many specific bodily coordinations and skills that he has already acquired. It is a period when toys and tools are of relatively little importance. Total bodily activities, rather than those requiring the finer muscular coordinations, are preferred. Vigorous activities are more effective in providing release of the child's abounding energy during this period: running, jumping, climbing, swimming, bicycle riding, and "stunting" of 'various kinds are typical activities, varying in form and detail, of course, from culture to culture.

The element of daring is a common feature, particularly in prepuberal play activities. Children during this period are concerned with little more than their status among their peers, which they enhance by achieving greater motor skills and by performing stunts and acts that require daring and courage. They challenge one another in such activities as walking high fences, climbing and swinging, performing on the high bar or the trapeze. As yet they are not at all concerned with the general problem of achieving status as an autonomous member of adult society.

There is a steady and rapid development of the more complex gross motor skills, particularly from age 9 years to 12 years. A great variety of skills involving agility and precise, large-scale muscular coordination are features of play at these ages.

Also characteristic of the play activities of school-age children are the more formal games involving gross motor skills. Gutteridge (1939) studied the development of the skills involved in such games as baseball. She found that although children begin practicing thtowing at age 2 or 3, it is not until they have reached 5½ or 6 years of age that the majority (75-85 percent) of children can throw a ball well. Throwing a ball, of course, is a complicated act. Gesell *et al.* (1940) analyzed the process of throwing:

> Throwing involves visual localization, stance, displacement of bodily mass, reaching, release, and restoration of static equilibrium. Skill in throwing a ball requires a fine sense of static and dynamic balance, accurate timing of delivery and release, good eye-hand coordination, and appropriate functioning of the fingers, as well as the arms, trunk, head, and legs, in controlling the trajectory of the ball.

It is during the preteen years, however, that throwing skill, along with catching, fielding, batting, and the other fine points of the game of baseball, are perfected. Team games requiring a rather high degree of motor skill constitute an important aspect of life during this period. Girls frequently participate with the boys in certain of these games, often acquiring skills comparable to those of boys. Govatos (1959), for example, found that on the average, at 10 years of age, boys did not differ significantly from girls in such motor skills as the jump and reach, standing broad jump, 25-yard dash, underhand ball throw for distance, and accuracy of ball throw. Only on such tests as involved superior strength in arms and legs, such as soccer kick for distance and ball throw for distance, were the boys found to be significantly superior to the girls.

It is during these preadolescent years, however, that sex cleavage generally is most pronounced. Blair and Burton (1951) describe this apparent sex antagonism:

> The apparent antagonism between boys and girls at this age is one of the most commonly observed characteristics of childhood. Parents and teachers have long recognized an inordinate amount of teasing between the sexes in the upper elementary grades as well as the almost complete exclusion of the opposite sex from the other's play groups....
>
> Although this teasing and antagonism usually appears about equally reciprocal, there is some indication that it

is more pronounced in boys. On the basis of her observation Zachry (1940) notes that girls of this age spend most of their time with girls of their own age. Often, however, they seem to be doing so less of choice than of necessity. If they assert that boys are horrid or nasty, their scorn does not always ring quite true. It is less convincing than the aloofness or teasing with which the young boy meets them; more often than not it is a mode of self defense or retaliation.

Boys, nevertheless, recognize playing skill, and in spite of any tendency they may show generally to reject girls from their company, they often welcome a good girl ballplayer on their team.

Fine Motor Coordinations

During these middle years, great strides are also made in the development of the more finely coordinated muscular skills. The schoolroom becomes an important segment of the child's environment. It is the business of the school, among other things, to see that the child acquires some degree of proficiency in such essential skills as reading, writing, music, and other arts.

The complicated process of learning to see objects and to discriminate fine differences, as we have seen, requires time. Reading from the printed page is a highly complex and difficult psychomotor skill, which is acquired in varying degrees of proficiency during elementary school years.

The companion skill of handwriting, likewise, is acquired at many levels of proficiency. Handwriting involves a complicated pattern of learned muscular coordinations. The muscles of the shoulder and wrist develop very rapidly during the early school years, but those of the fingers and hand used in writing develop more slowly. Writing is one of the skills in which both bodily maturation and practice are important factors in its achievement. It is but one of the many finely coordinated psychomotor skills achieved during this period.

Cognitive Functioning

During this seven or eight-year period the cognitive aspect of this total functional development, when focused upon specifically, is prodigious. By virtue of this total development the child continues to function cognitively at progressively higher levels of effectiveness. As we have already noted, he reaches school age (5 or 6 years) with a well-established ability to use symbols to represent familiar objects and situations.

A symbol is something which, in a particular culture, has come to stand for something else. The former "something," the signifier, may be a word heard or spoken, or in visual form a written or printed word, letter, number notation, and so on. The "something" for which the symbol stands, the significate, may be a familiar object or class of objects, situations, or relations. Symbols, then, are names or labels for concepts. They become invested with representational value by repeated association, in each case, with whatever they come to stand for in perceptual experience.

It will be recalled from our earlier discussion of the different "modes of representation" in thinking that the preverbal child often uses his own concrete acts to represent objects and their functions. His first signifiers are not words. They are his private symbols. "Shaking his legs represents the bassinet fringe; laying down his head, grasping the blanket, and sucking the thumb represent going to sleep; opening and closing his mouth represents opening and closing a match box".

But now at age 5 years many of the child's representations have become internalized. Sensory images can represent absent objects and past activities, and word symbols are used with facility as signifiers. This ability to manipulate word symbols is a feature of the school-age period (Piaget's period of concrete operations), which distinguishes it from the earlier preverbal period. The 5-year-old can now think with much greater facility.

But the 5-year-old thinks in concrete rather than in abstract terms. He is as yet incapable of analyzing and synthesizing, of rationally solving problems mentally. His thinking is in terms of representations of actual events and activities as he actually perceives them and participates in them.

1. The child is perceptually oriented; he makes judgments in terms of how things look to him. Piaget has shown that perceptual judgment enters into a child's thinking about space, time, number, and causality. It is only as the child goes beyond his perceptions to perform displacements upon the data in his mind that conservation appears.

2. The child has difficulty in realizing that an object can possess more than one property, and that multiplicative classifications are possible. The operation of combining elements to form a whole and then seeing a part in relation to the whole has not yet developed, and so hierarchical relationships cannot he mastered.

3. The child centers on one variable only, and usually the variable that stands out visually; he lacks the ability to coordinate variables.

This concreteness, this rigid adherence in his thinking to perceived sequences of events in experience, does not permit the 5-year-old to reverse the sequence. He cannot as yet "return to the point of origin" in his thinking. For example, if he is asked whether he has a brother he might answer "yes, his name is Jimmy." Then if asked whether Jimmy has a brother, he is likely to answer "no." His thinking progresses in an irreversible direction just as the events in his experience do. This is "preoperational thought."

As his cognitive development proceeds, however, he soon overcomes such rigid irreversibility, as well as a relative lack of "conservation" in his thinking. Even as a 7-year-old the child can "retrace" to some degree occurrences in the reverse direction in his thinking. He can, for example, judge the quantity of liquid to be unchanged when he sees that although the level of the liquid rises when poured into a thinner glass container, the width of the column decreases. His thinking is still concrete but more "operational" in Piaget's terms; he can manipulate representations—images and word symbols—of "concrete" objects and events in his thinking.

The child continues to live and function mentally largely in terms of actual objects, events, and activities throughout this period of "middle childhood," but with ever-increasing facility. These "concrete operations" establish a foundation for logical thinking, for "formal operations" as he approaches adolescence. But as a prepubescent he is not yet ready to be concerned with problems that require conceptual thinking.

Emotional and Social Adjustments

By the time a child is old enough to enter school his parents ordinarily have already done a fairly thorough job of "socializing" him, for better or for worse. As we have noted, his basic behaviour dispositions have become largely shaped. His individuality has thus become established. And even though he may have developed certain unpleasant or difficult emotional patterns and behavioural tendencies, as a general rule the 5 or 6-year-old is a relatively comfortable individual to live with.

It was noted above that with respect to one aspect of the child's physical maturation—the development of his reproductive system—the term "latency" is applicable. In other areas of development, both physical and psychological, progress is steady and substantial. However, there appears to be less surgency in his sex interests during this period as compared with either the earlier period or the puberal period to follow.

The later preschool years were described in the previous section as a time in which emotions are high pitched and the attachments to the parent of the opposite sex are often intense and presumed to be erotic in quality. According to theory, the whole situation, the Oedipal crisis, becomes quite intolerable, and relief from it comes through the mechanism of repression.

The child thus enters the school-age period relatively free from emotional conflict. He generally is quite content to conform to the behavioural standards set by his parents. His own level of competence as a person outside the home provides little basis for a feeling of importance. He feels warmly attached to his parents, his primary source of security. His sense of importance and of status is based upon his conceptions of and feeling toward his parents. As Ausubel (1954) put it, the child "revolves" about his parents. He adopts the role of a satellite to them.

> By doing this he acquires a derived status, which he enjoys vicariously by the mere fact of their [the parents] accepting and valuing him for himself, regardless of his competence or performance ability.

Such "satellization" apparently satisfies a real need at that stage of development, and the degree to which that need is satisfied depends very largely upon the general family situation and the attitudes and activities of the parents.

In any event, the release of the child into the world outside as he enters school brings changes in the general family routines and arrangements. The child is no longer completely under the control and care of his mother. The teacher now assumes part of that responsibility. With this change the parent often feels a sense of relief. Parents are likely also to begin to see their child as less in need of their care and direction. The tendency often is, to relinquish much of their former control over his activities, and the closeness of their relations with him is likely to decrease. The child spends more and more of his time with his age peers, both in school and out. Consequently, one of his major concerns gradually comes to be his ability to compete satisfactorily with his peers. This, of course, results in his acquisition of the many skills to which we earlier made reference. The child's sense of personal adequacy thus develops, and his need for the "satellizing" relationship with his parents diminishes. The whole trend is for the parents to pay less and less attention to their school child's activities, and hence to know him less intimately, and for the child to identify less closely with the parents and to be less interested in the parents daily activities. Family interaction thus tends to diminish or to cease.

In the meantime certain other behavioural and attitudinal changes are taking place in the youngster. Because of the usual lack of closeness between parent and child, these changes can come almost imperceptibly to the parent. Many important and desirable interests and competencies, of course, are likely to develop, but there is also likelihood of less socially desirable developmental trends. The prevalence of peer associations and peer-group activities results in "clubs," gang organizations, "secret societies," and the like, some of which may have constructive purposes and results, while others may result in different forms of juvenile delinquency.

●●

Index

T

U

V

W

Y